KRAKEN RISING

"The Admiral's Fight to Save Earth
from Catastrophe"

PETER KEATS

A LITTLE ABOUT ME...

I am someone who has been there, read tonnes of SciFi, and done it, but many years ago. Between those teens and now, there have been a lot of things in the way – life! Two kids, a second wife, and lots of heartbreaking news in the news every single day, it seems...

I exhausted most of what I might call the 'mainstream' SciFi writers a long time ago. Asimov, Clarke, Poul, Zelazny, Herbert, Biggle Jnr... oh, I don't know – the list is considerable.

Along the way, I suppose I might have considered growing up, but then thought better of it. How do you grow up reading Mad magazine? Dave Berg and his Lighter Side, the side cartoons by Sergio Aragones... all marvellous stuff.

I am originally from a town on the south coast in England, originally in Hampshire, but for some strange reason, the borders were switched, and it's now in Dorset. Go figure. I can't say that I enjoyed my school life, as it was brain-numbingly boring. Most of the time. We had a maths teacher who made my days worth living. Vic Loosemore. Quite an eccentric but brilliant man. Happier days...

I am, in spite of all of the modern-day madness that we are presented with every single damned day, an optimist. Man, as a species, will continue. The planet doesn't care. If Man vanished tonight, then the planet would not even know about our brief presence. It is up to Man to decide how we spend the next few decades. Do we bow down to threats and oppression from religion (I am an avowed atheist), or do we rise above the petty (and not so petty) squabbles of recent centuries and days?

I hope that we can rise above it all – all the trash that we are faced with every day. We are better than this, and we should be able to embrace the future rather than fear it.

Peter Keats, November 2015

TABLE OF CONTENTS

BATTLE STATIONS

The Admiral knew that he was in trouble – big trouble! The Galaxy-class starship came out of the jump gate in flames, from the front to the back, and was clearly struggling to survive. At a ridiculously high speed, it continued on what was very nearly a dead-hand course. Aboard the bridge, Admiral Avraham did his best to steer it, but he knew that there was very little control left – the damage was too severe. The jump gate had closed immediately behind them, which afforded them some security, as he knew that no one could be following them. It had been a random, mistaken dial-up, as the Admiral had just wanted to get back to Homeworld to report on this new alliance of their enemies, but since the rest of the crew on the bridge of the ship were all dead, he was fairly sure that he had been distracted and had made keying-in errors.

So, where the hell was he now? By Klono's claws, where had he jumped to? The computer systems in the bridge didn't help, but they were still scanning the surrounding planets and stars, trying to spot where they had jumped to. Was it a system that they had previously catalogued, or was it somewhere unknown? Right now, though, there were more important things on his mind. Having just escaped the battle scene, the ship was falling apart, with most of the systems down or failing. The Admiral, and by now what was left of the crew throughout the ship, knew the odds of survival were slim, but they had to take any chance that might still be on offer.

The ship had appeared above a blue-based planet, which, according to the few scanners that were still operating, held a compatible atmosphere. It wouldn't kill them! If they could just get down in one piece, then perhaps what was left of the crew would not die today. Since there were so many things that were wrong with the ship, the Admiral had little control over the direction or speed, and the planet was right in the flight path! He was hoping that enough of the controls would stay together long enough and that some of the crew would survive the descent. He knew that various parts of the ship were totally wrecked, and thousands were already dead. It would be an interesting next few minutes....

The jump gate had been created halfway between this planet and its only moon, positioned a long way below the line joining them. The control of the ship was getting worse by the second, and the Admiral knew only too well that the survival of what was left of his crew was getting more doubtful. He had to put a last message out before hitting the atmosphere of this strange new world.

"Attention to all of the crew. This is the Admiral. All hands, prepare to abandon ship. Try to take anyone who is injured as well. Find any escape pod you can and use it! We are about to enter this planet's atmosphere, and all that I can guarantee you is a very bumpy ride. The pods should allow you some freedom of travel, so I hope that you won't just end up in the water. If you have the time, try to lock on with some of the other pods so that they follow you. We don't know what may be available on this planet. You only have a couple of minutes, but try to salvage anything that you can, and take it with you! Look for land, and perhaps some of us will make it. I will just say that it has been my honour to work with you all, and I hope to see some of you when we get down to the planet, about which, at this time, we know absolutely nothing. Oh, apart from the atmosphere, it will not kill you! "All fighter pilots, you are free to launch when you can." He paused and then added, "And may Klono have mercy on all of our souls!" The reference to the mythical deity called Klono was heard by the small number of fighter pilots who were still in control of a flyable scramjet, and they all laughed – and

2

then launched. They all knew that they needed to put some distance between the ship and themselves, or they would perish on the entry into the atmosphere. Unlike the mothership, the fighters were omni-environmental, and they could cope with almost anything, whether in an atmosphere or in space. The Admiral hoped that once they had landed, the fighter pilots would be able to stay hidden for a few days, until things had been sorted out.

The ship came in fast from the south, arching over the sky, the controls almost completely gone. It was complete luck that the ship didn't bounce back off the atmosphere and get lost back into space or fly into the sun. The Admiral put every retro engine that he could get to work into operation to try to slow the descent, but the damage was too bad for it to have much effect. The ship came in, literally falling apart across thousands of miles of the planet below. Everywhere the ship passed, bits of it fell away, dropping to the ground with great destructive force. Although he didn't know it at the time, a few hundred million people on the planet died that day, either from direct impact of the debris or from secondary impact damage over the next few days. The route of the ship changed as parts of it exploded off. When the engines blew up, the ship was slightly nudged this way or that way.

The population of the planet watched both in fascination and horror as the flaming ship fell past them. At that time, nobody knew what it was that had shown up in the sky. Many people were under the flight path, but many more were not. They all went mad with cameras and videophones, recording what they thought was a large meteor coming down, not realising the chaos and destruction that would follow just seconds behind it.

On the final part of the approach, with all of the controls by now completely destroyed, the Admiral could do nothing more. The ship was now a falling piece of disintegrating junk, with no computer mind left to do any thinking. He struggled to his own escape pod, knowing that there was no one else alive on or close to the bridge. Although the pod was designed for up to four people, he knew that he was alone for this trip. He set up to slave in as many other pods as he could automatically find and punched the launch button, hoping to get clear of the final destruction, and as he flew out of the ship, he could see the extent of the damage. The ship was in flames from front to back, with a trail being left in the atmosphere for thousands of miles. He cursed, wondering what this damage was doing to whatever the population of this planet consisted of. He was pleased to see another handful of pods launch at the same time, their controls following whatever his pod got up to.

The escape pods had been designed many years ago, and little had changed in that time. The fuel quality had improved over time, but to look at and to handle, they were pretty much the same as they ever were. They were an inverted pear-drop shape, with the large part at the top. Because of their shape, they were good for re-entry into most atmospheres, and their storage capacity was surprisingly large. They had extendable legs for landing and a long and very strong ramp for entry and exit. Each stood around sixty feet tall, with three floors for people and the rest taken over by engines and computer storage.

His escape pod was semi-intelligent, and it turned and flew away from the likely impact area. The others followed behind it. He let the pod go, too worn out to care. All he wanted to do was to get to the ground in one piece! Although he knew that he shouldn't, he felt that he had to close his eyes, just for a second, to rest. He immediately fell asleep and was woken just moments later by the alarms sounding as the pod was coming in for a landing. He panicked for a few seconds, unsure of where he was, and then he remembered. He remembered too much, and it sent a shiver down his spine. All of those ships, all of those brave people, all lost to the void of space. He couldn't guess at the numbers, but his brain did some rough calculations with six thousand people on each Galaxy Class Battleship, which equalled sixty thousand people, all now gone. The other crafts were all smaller, but there had been hundreds of them. He stopped trying to count. It was too depressing.

He could see that the instruments in the pod were all functioning, so he started to read them to find out what was going on. They showed that there were five more pods all following his, but none of them showed any life signs. The atmosphere was slightly different from Homeworld, but close enough. Remarkably close, actually – just a two per cent change in the oxygen and nitrogen mix. He and the remnants of his crew would survive today after all! If the pod could land properly, then the Admiral would not die – not today, anyway. He looked out at this new world, which was likely to become his home for the rest of his days. Was there intelligent life here, he wondered? What level would they be at? Sticks and stones or further along the evolutionary ladder? Leave it, he told himself. You'll find out soon enough.

The pod did a good job of coming in low and under control, and the others just followed the pattern. His pod didn't rush but looked around for a safe spot to land. It didn't take long to identify a flat area, and the pod was moving carefully towards it. This time gave the Admiral a chance to inspect the pod instruments. There was a lot being shown; he just hadn't paid attention. There was electrical traffic from various locations on this planet – there is a population, and they at least have a radio! That might be a good sign.

He then checked the pod log for details of what had happened during the descent and was horrified by what he saw. Because he was the Admiral, his pod had a much bigger storage capacity and was constantly updated with telemetry from the entire fleet. He could identify all of the actions that had taken place.

He saw the long trail of destruction in the atmosphere as the main ship, three miles long and half a mile wide, had cut through the air, a thing it had never been designed to do. Quite a few years ago, he had been present for a large part of the assembly of this ship, some two hundred thousand miles from Homeworld. It had been put together in a space dock, a huge area that was designed to do exactly this. Assemble millions of parts together until you have a jump-gate-capable starship. Because it had been put together so long ago – in Earth terms, thirty years – and because even now the technology on Homeworld was still changing, still advancing, then the scramjets were a totally different thing altogether. Their engines were designed more efficiently, and they could achieve speeds that were frightening to some. Because every craft had artificial gravity, it meant that the scramjets were travelling at thousands of miles per hour within seconds from launch. The Starships were just huge and heavy, with hundreds of super-powerful engines. It took time to start them and to steer them, and if they achieved any speed, it took time to stop them, too!

He could see every detail as sections of the ship just fell off, pulled away by the enormous drag of the air. The largest parts, he knew, were many tonnes in weight. He shuddered as he thought of the damage they would cause on impact with the ground. The path had included a long flight over one of their oceans, and he could see the tidal waves caused by landing debris. Many low-lying islands had already been flattened, and the tidal waves were still continuing in all directions.

He bit his tongue and cursed, watching in frustration as the instruments told him of the chaos that his arrival had caused. Oh, hell, if there is intelligent life here, I've just wiped out most of it! Whatever is left will not be very friendly when they find me. If they find me, he added…

He couldn't know it at that time, but during the passing of the ship and its ultimate demise in the hostile region known as the Himalayas, over two hundred million people died. The damage caused during the following days would more than double that number, and over the following days and weeks, the figure moved close to two billion. Many millions were quickly extinguished, burnt or crushed as debris flew from the ship and landed everywhere. A lot of major cities were affected. Buenos Aires and São Paulo were both flattened, as were many smaller towns in between, when the first batch of the engines dropped away, landing with the force of nuclear explosions. It wasn't just the debris itself but the fuel. Once any of the one hundred engines had fallen off, it took its fuel supply with it, sometimes exploding high in the air or sometimes crashing into the land and then exploding. Each time, it raised a small mushroom-shaped cloud, which then started to slowly spread due to the weather conditions. Across the Atlantic, the ship flew, dropping hundreds of tonnes as it went. Ships that were caught by the falling debris simply vanished. It went over West Africa, passing over Sierra Leone and Burkina Faso, and on into Niger, Chad and Sudan. Across the Red Sea into central Saudi Arabia, southern Iran and then Pakistan, flying over the top of Nepal and then crashing down into a desolate part of Tibet.

All along the route, there was an endless catalogue of secondary damage. The engines didn't just drop down; sometimes they exploded out from the ship, going left and right. In the early part of the descent, this meant that they flew dozens of miles to either side of the flight path. Flying over Iran, an engine blew apart from the ship and landed at a shallow angle in Namak Lake, just outside Qom. This engine then exploded at a depth that caused a huge wall of water to fly out from the lake and drown out the southern part of Tehran.

The list of damages just grew by the minute.

A STIRRING IN THE DEEP

The world had thousands of earthquake measuring devices in every corner. They all went mad. Every single one of them went off the chart – those devices that survived the descent of the ship, that is. There were hundreds of research stations set up everywhere, and the attending staff at first just couldn't figure out what was going on. Those stations in Antarctica registered more in the first few hours than they had in the decades that had gone before. All the stations that were anywhere close to the intersections of the great plates registered that this is it, this is the big one. Thousands – millions! – went crazy in San Francisco, thinking that the Great Quake of 1906 was about to be surpassed. The shock waves from the explosions went everywhere – across the globe and into it. Fractures started to appear in many of the tectonic plates. The damage was immense, and it grew by the minute!

Somewhere deep, buried under many yards of sand and dirt in the Mariana Trench, in the Challenger Deep, a strange shape did some thinking. It had lain there for an unknown number of thousands of years, but with all of this sudden noise, it just had to wake up. It shook itself and said the modern equivalent of 'Just what is going on here?' To call it 'him' or 'her' was irrelevant – it was above that station in life from the moment that it had been created.

Without moving from the spot, it sent out sensory feelers. It looked around and saw that this planet had changed dramatically since the last time it had actually looked at it. It scanned the skies and tried to work out where it was in the galaxy by using the positions of the stars. It only took a few moments, and then it had its answer. It had lain where it was for close to two hundred thousand

9

years. No wonder this planet looked so different! It focused on what it could find that would be considered of use and then did something that it hadn't done in a great many thousands of years. It moved. It needed to find out what had just happened on the planet that had made all of this noise and what had happened to wake it from its long slumber!

It was a creature as old as time itself, with the same terms of reference. What did it matter if this form of life called Man had appeared during a few days, by its own time, when that time frame said that this planet had been around for a ridiculous number of centuries – millennia, actually? What did it matter about the current life forms that considered themselves to be the highest form of intellect on this backward sphere, in some easily forgettable star system?

It decided that it would investigate a number of things. Who was it that considered themselves to be the highest form of so-called intellect on this planet, and who had made all of that noise, waking him from a long sleep? He was not a happy person when he decided to make the first move. *Someone is going to have a bad day for waking me up from my peace and tranquillity!*

THE ADMIRAL

CHECKS THE WORLD OUT

The admiral started to look at the details on the available systems as the pod came in to land. He was pleased to see the other pods follow in on a path that put them all very close to each other. He immediately picked up on the signals coming from this new planet. These people didn't just have a radio; they were transmitting pictures as well! That meant quite a sophisticated level of technology, which the Admiral found encouraging. He then identified a network of electronics that the population called the Internet. He logged into this, and instantly the pod was fed terabytes of data, everything that there was to know about this world. He activated the homing beacon on the pod and wondered how many others might have made it down and where they would be – the long route that the ship had taken coming down meant that the others could be on the other side of the planet! How would he get to see them again?

He monitored the airwaves and identified dozens of different languages. Each region seemed to have its own vocalisation, which he found to be very worrying. That meant splintered societies, since the unification was the only way to make things consistent. On Homeworld, the Unification had happened centuries ago. True, it had taken many centuries *before* it had happened, but the progress since Unification was enormous. They had shot through centuries of development in just decades, discovering all of the science-based cures for ninety per cent of all their diseases and founding their space-faring technology. Jump gates and huge spaceships were now common on Homeworld. So, too, was matter transmitting, albeit with a limited range, but it was more than enough for their own planet. Fancy a holiday away, on the other side of the world? Take a MatMit and be there in ten minutes. Matter transmission was common.

The Admiral wasn't stupid – in fact, he was quite the opposite. He was one of the most intelligent people that Homeworld had to offer. He was good at almost everything he set his mind to. He had been a natural for the post of captain of a new Galaxy-class starship, and it took less than a year for the Forces Commission to realise that he was capable of more. He had become Fleet Leader next, with the job of going against the pirates and the invading aliens. He had five years of success, solving many battles before they had really begun and winning many fights that the opponents were to regret ever starting. He became something of a legend among the crews of the fleet. He became infamous to all of the pirates and the aliens and, along the way, rose to be Admiral of the Western Fleet.

The effect of this was to catch up with him on this last fight – the pirates and the aliens had joined forces, with the one task of eliminating this Hero of the Fleet. The battle had been short and very bloody. For a few minutes, there were lasers and photon torpedoes fired all over the place, and ship after ship had exploded. Just when they had whittled each other down to a handful, the pirates had launched their reserve fleet, the aliens, hidden on the other side of the planet's moon. He hadn't planned for something like that, something so utterly incomprehensible.

It was enough to turn the tide. The Admiral saw the last few ships take massive damage and vanish in balls of fire. He had realised that he was set up to fail and invoked a last desperate act. He had enough space left to create chaos all around him. He launched a full spread of mines – well over one thousand of them –

and then punched up for the Jump Gate. The mines were computer-controlled and would seek out anything that was not classed as friendly. As he flew into the gate, the first of the mines found their targets, and he knew that the enemy was going to suffer at least a few more casualties.

After all of that activity, to land the pod he was in and the others was almost an anticlimax. The pods had landed safely, and the admiral relaxed a little and assessed his situation. He did it calmly, with all of his professional training coming in, so that he maximised his potential for survival. One thought kept going through his mind: What are my chances?

Just at that moment, the pod systems alerted him – they had done their calculations and now knew where he had landed! Well, they sort of knew where he had landed… He looked at the readouts and was shocked to see that the pod had worked out that he was now about 15 light-years off course. His keying in had gone completely wrong and sent the whole ship in the wrong direction by 15 light-years! His home was now some far distant place, and he doubted that he would ever see it again. He mapped the results into the Internet systems and located his home world as Epsilon Eridani. He knew that his 'home' was the fourth planet, something perhaps doubles the size of this new planet that he had just crashed into. Because of that, he knew that he and his people were much stronger than anything that this planet could have developed. His race had developed to be super-strong – he would have to be careful if he met anyone, as he would crush them like glass. During the years of his service, he and many others had also been surgically altered to increase their strength and overall speed. This was a real risk to anything that came into his path.

He looked at the instruments, going through them in a prepared order. His fuel was surprisingly high still, and he had perhaps another few hours of flight time left. The other pods all showed a similar fuel level, which he found reassuring. If he needed to, then he could take some of the fuel from one of the other pods to keep his pod going for a little longer. The outside air was easily breathable, even if it was perhaps a bit cold at this location. He checked the radio and visual transmissions, hooking in to all of the local bands. He left them running in the background, so he heard the chatter on the airwaves. It would take him another few hours, but then he would probably be able to pass as a local in a rudimentary conversation. His brain worked in many ways, and picking up new languages was just one of them.

He stayed there for a few hours and slept when he could. He was clearly in an isolated area, as no one showed up on the scanners within twenty miles. He was in no rush to meet any of them. He needed time to try to understand something of this world that he had literally crashed into. The radio signals were going mad with everyone screaming at everyone else about what had just happened. He slowly came to understand what was being said and was saddened by what he heard. The casualty numbers were huge and growing every hour. He identified other frequencies and switched to other transmission stations that offered global coverage. He found out about the chaos in South America, the millions confirmed dead in just that area, the unconfirmed losses across the ocean, and the massive numbers of dead across Africa and into the Middle East.

He was switching through all of the channels and was shocked when he heard something that was very close to Homeworld speech. 'This is strange,' he thought. It is not quite the same, but very close. He identified it as something called Hebrew, coming from a country called Israel. He ran through the maps showing on the screens and found Israel very quickly. He then spent the next few hours listening to their broadcasts, as he found that he could understand almost all of them. After three hours, he could pass as a native, he was sure.

He heard of the effects that the descent of the ship had caused, the trails of destruction, how some countries had almost been wiped off the map, and how emergency services were just not able to cope. Nobody had ever planned for anything on this scale!

He was alarmed by what he heard, as the Israeli stations broadcast speeches from other countries, supplying a Hebrew translation alongside. There was something called the United States of America, something called Russia, a Europe, a Japan, a China, and many other countries around Israel that were all yelling and screaming about what had happened. He began to fill in some more details, fearing where this might lead. He began to identify what had taken place, and he didn't like any of it.

On approaching the atmosphere, high above some South American countries, a city called Buenos Aires had taken a huge casualty rate as large parts of the ships fell away, including the first engines. As the engines had shot down, they had exploded. The overall effect was as if a nuclear explosion had taken place at an altitude of five miles. The city, and much of the surrounding area, was flattened within seconds. Anyone within that blast area perished. In the radius around the area, there were at least ten commercial airlines, some on approach paths to Buenos Aires airport. Half of them vanished, and those that survived suddenly had no airport to land at. The pressure created by the explosion caused a tidal wave that went north from the city, crashing through everything in its path, finally demolishing Gualeguaychu. Another branch of the tidal wave went up the Rio Paraná de las Palmas to Zárate and further, taking out the nuclear facilities at Atucha, nearly forty miles inland.

As the ship continued along its doomed flight path, the casualty list just worsened. São Paulo, Rio de Janeiro, and all of the smaller towns under the ship had damage. Many small towns just vanished, their population vaporised, along with all of the buildings. According to the news, he heard that a large statue of someone famous in popular culture had also been destroyed in Rio. He checked out what they meant by this 'Christ' character and was worried by what he found. They still had religion on this planet? He swore again, something he rarely did.

He found out that the flight path had continued across a large body of water that was called the Atlantic Ocean, and the ship dropped more debris on the way. Large cruise liners and oil tankers had all vanished, and the tidal waves were still radiating out, causing yet more damage. Those countries at the top of South America had avoided any debris, but then the tidal waves hit, wiping out whole coastal areas. The same was true in the Caribbean, and Florida and the Eastern Seaboard of the United States of America then took a massive hit. The other countries of Western Africa were also slammed, the coastal areas suffering deaths in the hundreds of thousands. As the ship had flown over the West of Africa, it had caused total devastation in many of the areas. The damage caused in South America was repeated here in Africa, down as far as South Africa and up as far as something called Great Britain.

On the ship flew, oblivious to the catastrophes being caused below it. When it crossed the Red Sea, by sheer chance, in the tiniest of windows, it dropped nothing. Nothing hit the water. Places called Sharm el-Sheikh, Eilat, and Suez escaped devastation by a billion-to-one set of odds. Unfortunately, the ship then started dropping again, in time to hit the first bit of land, somewhere called Saudi Arabia. A place called Mecca was made to vanish; everything was lost forever. It flew on and continued to drop on Riyadh and over the Persian Gulf. While crossing this water, the ship was in its death throes, and hundreds of tonnes dropped, causing massive tidal waves that flattened everything they landed on. The western coast of Iran was drowned, as were Bahrain, Qatar, Kuwait and the United Arab Emirates. Dubai, that pearl of the Middle East, simply vanished under a two-hundred-foot wall of water, all of the ridiculously tall towers collapsing into the floodwaters. The southern shores of Iraq went the same way. The southern movement of the tidal waves moved out at great speed from the Persian Gulf and on, hitting Pakistan, drowning Karachi and all of the coastal towns, and then it hit western India, drowning thousands. Because of the direction of the water, the Gulf of Kutch trapped millions of gallons of water, effectively reducing the towns around it to rubble. The ship still flew on, passing over Pakistan's northern territories, causing more disasters as it went.

He immediately searched this utility that the population of this planet had, which they called the 'Internet', to find out about the current state of play of this 'religion' and became very worried by what he discovered. Religion was splintered, and the languages are splintered – I wouldn't be surprised if this planet were still having wars on a regular basis. Just for the hell of it, he started a subroutine to search for 'WAR' while he continued to listen to the broadcasts.

As he listened, he thought that he found a common theme among various groups. The main story was that nobody had a clue as to what had happened. That became painfully obvious as he listened to the most obscure nonsense being spouted as 'fact'. There were these 'religious leaders' in the USA who uttered that it was the 'End of Days', whatever that was. Others quoted that the tragedy of

17

today was because the USA had recently accepted that same-sex marriage was now to be considered legitimate. *Klono's claws – we've had that for centuries!*

There were clear signs that something had happened on a global scale, but not a single person was making sense. It was also clear that this country, Israel, was being lined up by most of its neighbours as some kind of fall guy. In its simplest form, it came down to, 'We have no idea what happened, but we are blaming Israel anyway.' It really was that simple.

A thought ran through his very busy mind – just what are these people talking about? They have advanced technologies, even, he noted, atomic weaponry, and yet they still resort to the prehistoric stupidity of blaming the other guy? There are some seriously screwed up people on this planet, and I need to be very careful where I walk, because I need to avoid these fools. *There must be some sane people around here, somewhere? Please?*

He scanned the frequencies, hoping that some of the other pods had landed safely. He wasn't keen to think about what he would do if he were the only survivor. He had a few responses, and the other pods sent their planetary coordinates. There was no one near him for thousands of miles, so effectively he was on his own. Nobody opened a voice communication, and he didn't suggest that they should. It was better to be quiet for a while, until people had worked out what being a person on this planet actually meant.

He played through the options that he could identify. What to do? He couldn't stay here, in the open and very exposed, but he wondered where he could move to. If he were to try to get to this country, Israel, what would that gain him? Would they be able to listen to him as he tried to explain? What obstacles were along the way? The pod was showing enough fuel to fly for some time yet, if he was economical. That meant that he couldn't afford to get into many arguments along the way, but just to be safe, he went to the closest of the other pods and removed a fuel cell. That doubled his available travel time. While he was out, he checked in on all of the other pods to see that they were usable and what supplies they might have. He was planning ahead, wondering how long his food would last.

He looked at the maps and tried to plan a few different routes, some direct and some that had to go around a few countries and cities that he had already marked as 'potentially hostile'. Given the method by which he had arrived, he reckoned that most of them would be hostile anyway, so he needed to be very careful for the next few days.

An alarm on the pod rang out, and when he looked at his systems, he could see three fast-flying devices coming towards him. He was very exposed where he was and didn't want to meet anyone from the planet yet, so he looked at the mountains around him and could see a range of overhanging rocks which would afford some cover if he could get there in time. He fired up the pod, on minimal power, and slowly lifted off. The other slave pods followed close behind. He was watching the arrival of these flying devices, what the population called 'planes', and he tried to judge the distance to the mountains. He pushed the pods a little faster and arrived under a huge outcrop just as the leading plane flew overhead. He hoped that he hadn't been spotted.

He was scanning the airwaves, looking for any conversation that the pilots may be having, but found only static. Perhaps they were on radio silence?

He kept the pod's system operating at low power and could see that the planes passed over another couple of times, and then they left. They must have tracked me down somehow, but they only have a rough idea of where I landed. That means that a land-based team will be arriving soon, he thought, the professional soldier in him working through the details. *That's what I would be doing.*

Since it was now getting dark, he thought he had better make some plans to move to a safer location, but where was there one? Since he didn't really know the details of this planet, he could leave his current position and make his situation even worse. He knew that somewhere there would be ground troops making their way towards his estimated position, so he needed to put some distance between himself and them. The pods were good for a few hours of travel, but he doubted that he could find any fuel on this planet – the technology was sophisticated, but only up to a certain level. He thought that he needed a technology jump of a few years before the fuel that he wanted would be available. Although he didn't know how he was going to do it, he realised that he had to make his way to this country called Israel. Once there, he would have to revise the situation and make new plans accordingly.

He looked at the options and explored the direct flight path first. That meant flying back through Northern India, some of the parts of Pakistan that had been obliterated by the ship on the descent, some of the central or perhaps southern parts of Afghanistan, into central Iran, and then across the Persian Gulf. Since the news channels on this 'Internet' were now starting to list the huge damage caused by the passing of the ship, he doubted that there would be much interference from any of the population – those that might have survived the passing of the ship, that is.

He looked at the map and wondered if he was safer going over the Persian Gulf into Saudi Arabia, across that land and then arriving at the southern tip of Israel, a place called Eilat. He hoped that, should he arrive at Eilat, he could make his way north to where their power base was held. If he managed that, then he would investigate, find some people whom he could talk to, and hope that they would listen. He knew that half of the planet must be after his blood because of the scale of the slaughter caused by the ship, even though they had no real explanation for what had happened. The news media was close to meltdown with conspiracy stories and terrorist threats (whatever they were), so he knew that above all else, he would have to be very cautious in everything that he did from this moment on.

He was utterly drained. Exhausted. He had been awake for four days, planning and replanning all of those battle sequences, going over everything that he could think of, either by himself or with his teams of officers. How many of them had survived, he wondered? He powered down the main systems of the pods, leaving only the proximity alarms. If anything came within thirty miles, then he would know about it. Otherwise, he slept. His plan was to wait until it was fully dark and then use the systems on the pod to fly towards Israel. He may not make it in one attempt, so he was keen to find stop-off points if he could.

MORE STIRRINGS FROM THE DEEP

Although his pod had the most sophisticated technology that there was on the planet, because there was so much going on, and his instruments were showing him so much detail, he missed the tiny entry that showed up, saying that just south of Japan, there were indications of some seismic activity deep in the Marianas Trench. He needn't have felt guilty, though, because everyone else on the planet missed it! Something was happening around that area, but everyone was looking elsewhere.

THE ADMIRAL

RETRACES HIS STEPS

He awoke, uncertain of where he was. It was dark in the pod, and he only had night lights. He had been dreaming of the last day and all of those fights. He was in the midst of the battle again, running through all of the instructions that had been issued, those that had worked, and those that had not. He saw only the dead, numbered in their tens of thousands… He shook it off and focused on the things that he needed to do now.

He went through the pod's systems, making sure that he and his location were secure, and then he set his mind to the next part of the journey on this new planet. He caught up with the news, and he realised that his earlier search on 'war' had returned ages ago, with millions of references. He started to look at it and was horrified by the overall summaries. The history of this planet, and that of its population, goes on for centuries, and they never stop having wars! He identified half a dozen current battles, still causing grief and heartache, in and around the Middle East, as it was known. He moved the scope out to find that North America had its own history of wars, and South America, too. So many of the population got caught up with other people's battles, he thought.

He knew that he had to start his new journey, and now. He activated the barest minimum of the pod's flight capabilities and slowly moved off, the others following close behind. He went in the general direction of Pakistan, intending to backtrack through the chaos that his ship had brought to that area. If he were lucky, he would find his way across part of Iran up to the Persian Gulf.

Depending on when he got there, he would decide whether to continue or to submerge under the water.

He immediately had some logistical problems. This was a very mountainous area, and his choices were to stay low and try to avoid the mountains themselves but spend a lot on fuel, or should he rise up a mile or two and go in a straight line towards his destination? He knew that he did not have an endless supply of fuel, so he climbed, very slowly, above the mountains. He was monitoring all the local bands for any indication that he had somehow been tracked, but they stayed silent. He could slowly increase his speed, again saving some of that fuel, and he would have to see how far he could get. The pod systems showed that it was over two thousand miles to the Persian Gulf. He would need a good run to get there in one night, so it's all or nothing, he thought to himself.

He advanced the acceleration until the pods were travelling at five hundred miles an hour. Very fast in the daytime, but more frightening at night when you are flying by the instruments alone. He had done this before, in training, but this was the first time in 'real time', so to speak. This wasn't training; this was seriously real. Make a mistake now and it could end you. There is no 'restart the mission' option here!

He flew through Pakistan, seeing on the night vision from the pod that it was chaos everywhere. There was no end to the number of towns and villages that had been flattened or blown apart by the parts of the ship coming down. Thousands died here, perhaps millions. The emergency services were simply not prepared to cater to this scale of disaster, and already calls were going out to the international community for any help that could be offered. He searched the South American radio frequencies and found that the surrounding countries, those that had not been trampled by the passing of the ship, were all offering what they could. It did look as though the 'international assistance' was coming through, but nowhere could they have anticipated the scale of this disaster. It had crossed a large part of the planet and affected billions of people. Those left, those on the outside of the problem, would do what they could, but even that help had its limits. The Admiral knew this and appreciated the issues that he had brought with him. The fallout from the passage of the ship would be immense. Thousands of plumes of smoke flew upwards and, over the first few hours, started to spread outwards. Those countries and people who were not immediately under the flight path came to realise that the 'collateral damage' was going to be everywhere. Nowhere on the planet was safe. It was only a matter of time before the air currents carried the smoke and dirt across the whole planet, effectively shutting out the sun. To many, the Nevil Shute book 'On the Beach' came to mind. That was a story of a nuclear war, which took out most of the Northern Hemisphere. Those in the southern hemisphere survived, but only for a short while, as the radioactive clouds spread further and further over the planet. Australia was one of the last to give in to the inevitable demise, but slowly, the whole human population was eliminated.

Such was the scale of the disasters that the Admiral was seeing; he knew that this planet was doomed unless he could activate some of his established technology to try to 'clean' the air. This was a remote possibility, but only if he could find the right people who could help him. They needed to be able to accept him for what he was and understand that time was running out for everyone. He hoped that he could find such people in Israel.

He moved on, reaching the end of Pakistan and moving into the country called Iran. Here, he plotted to follow the path of the incoming ship, and it worked well. There was other air traffic around him at times, but they were so busy ferrying people or supplies around that one more high-flying craft in the skies was just ignored.

He kept in a straight line across Iran, aiming for the northern part of the Persian Gulf. There, he hoped to have a little time to skirt around and see if he could find a safe haven for the daylight hours.

He was not interrupted at all, and he flew over Bandar Bushehr as he saw that light was starting to show in the sky. Time to take a break, he thought. He flew out into the Gulf, and at a point he estimated to be halfway from either side, he slowly dropped the pods. He could see that the damage from the ship was still being assessed by all of the countries affected. The news outlets that he could see and hear were all still going mad about this 'whatever it was'.

Kuwait City and most of Kuwait had vanished, just torn to pieces by tidal waves. Bahrain and Qatar were similarly affected; both had most buildings destroyed up to five miles from the coast. In the United Arab Emirates, little had survived along the coast, and in the Strait of Hormuz, little was still breathing. Khasab and the entire peninsula had vanished, washed away.

He could see that in Iraq, the city called Basra had been washed away, due entirely to the river deltas. When the tidal waves hit, they all funnelled up the rivers, wiping out anything in their path. The waves travelled for hundreds of miles, and town after town was just smashed out of existence.

The Admiral saw all of this and was close to tears, but his professionalism came through. I'm no use to myself if I surrender to this, and I am no use to anyone else, either.

He let the pods find the bottom of the Persian Gulf, and he closed down all of the systems that he could. He didn't want to be picked up for some accidental random transmission. Once he had done all of that, he slept. He was trained well enough to know that if he needed sleep, and he certainly did, then he had to grab it when it was on offer. Even if he only managed three or four hours, he must use them wisely.

He slept a deep, exhausted sleep. Again, because of his training, he woke some four hours later, feeling a little refreshed, but knowing that what he really needed was a week off. He chuckled, knowing that a week off was simply not going to happen. Maybe next week?

He activated what he could of the pod systems. He could see that there was a lot of water traffic, but all of it was going over or around him, showing no signs that he should worry.

He wondered what he should do next. He had been very lucky on the first part of the journey. Could he be that lucky for this second part? Could he aim to go straight across the Saudi Arabian and Iraqi border, aiming for the northern part of Israel?

He told himself to stop being so silly – stick with the original plan! He looked at the maps, tracking back through Saudi Arabia, following the destruction trail from when the ship came down. This would put him into the Red Sea area, where he could turn right and travel up to Eilat. He hoped. Let me get there first, he mused.

He continued to monitor everything he could and noticed his fuel levels were still very good. He waited until it was dark and lifted the pod slowly off the Gulf bottom. He made a decision to leave the other pods where they were, safe from prying eyes. He could activate them remotely if he needed to. He moved slowly south, heading towards Dammam, which he knew to have been badly hit. He rose above the water, five miles off the coast, and reassessed his situation.

He used the powerful viewing facilities within the pod to look at the coast and was shocked to see the damage. Every building was ripped to shreds, as the water, when it hit, just took all before it. He checked through the maps and could see that it was a straight-line journey back through Riyadh and then on to where this city called Mecca used to be. When he gets to Jeddah, he should be able to make it up to Eilat and then on to Jerusalem – he hoped.

He set the pod to rise above the land, to an altitude of two miles, and then he flew back through the chaos. Not too fast and not too slow, so he would not call attention to himself. He made it all the way to Riyadh before he noticed the air traffic all around him. He had been so keen to see the damage trails that he had forgotten to look up, around him. He was being buzzed by a pair of Army helicopters, which were making noises on the airwaves. He realised what the language was and gave a fair representation of "Sorry, people, but I have places to be." And then accelerated away, reaching one thousand miles an hour within two minutes. That had left the city and the helicopters far behind but had also started to drain his fuel reserves. He knew that he would be at the coast in a little over half an hour, so he smoothed off the acceleration and left the pod to run at its current high speed.

THE ADMIRAL

GETS NOTICED

Due to the high speed of the pod, he now started to alert other people. Various radar stations were still operational, not the least of which were various locations within Israel. He had gathered that there was a lot of technical secrecy involved, so he wasn't surprised when the pod alerted him to the fact that they were being tracked. He brought the pod into a right turn and arched away from Mecca – or what was left of it – and raced ahead directly towards Eilat.

It only took a few seconds, but then voices started being heard in the pod, in Arabic. "Unidentified aircraft on rapid approach to Eilat. We are tracking you, and if you don't respond, then we will assume that you are hostile, and we will defend ourselves with appropriate action."

That was it – no messing, no playing around. Tell us who you are, or we will get really annoyed! He didn't know in detail, but he was pretty sure that they had various forms of surface-to-air missiles, and then he could see a flight of three very fast planes coming down towards him. Who do they belong to – Israel or someone else?

They were on him and past him within seconds – they were going at twice the speed of sound on this planet! That's impressive, he thought. More voices came into the pod. "This is the Imperial Jordanian Air Force. We have seen you, and we do not identify you. Please tell us who you are and why you are in this airspace."

Again, no playing about. I'm dealing with professionals here. He could tell that from the voice. It's a quality that you have, that you develop, when you get to be that good at your job. He smiled. He would hate to have to hurt any of them. He looked at the map of the area and realised that the planes should have been from Saudi Arabia, but he wondered if that country had been hit so badly by the descent of the ship that their air force was all but useless.

He opened a channel to the Jordanian Air Force planes and, at the same time, opened a channel to Israel. In what he hoped passed for fluent Arabic, he said, "Gentlemen, please listen. My name is roughly translated into your language as Admiral Avraham Avraham. I have no wish to be regarded as hostile, so please do not make any hostile moves towards me. Like you, I am a professional soldier, but I am not from around here. I have travelled a long way to be here."

There was plane-to-plane chatter and then plane-to-ground base chatter while the pilots tried to make out just what was happening. They had seen the pod; it had been filmed when the planes went by, and the ground crews were even now poring over the photographic details. All the while that this was going on, he was approaching Israel. He spoke in Hebrew. "Israel, I would like to come in to land at your Eilat airport. I realise that you have some military personnel there, but please do not make any aggressive moves. I say again that I am not here to harm anyone."

He flew over Tabuk even as he saw the planes turn at high pressure. Those are good pilots, he said to himself. He switched easily between Arabic and Hebrew in the next few conversations. "Gentlemen, to the pilots in the planes. I hope you did not hurt yourself with those tight turns. That was impressive flying. If you wish, you may pull up alongside me, but please do not make any moves against me. I am armed and can defend myself if I have to."

He knew that it was a half-lie, but since he hoped that the pilots would not be able to tell whether he was armed or not, it was a risk that he had to take. Owing to the nature of the pure fuel that the pods ran on, he knew that should one ever explode, it would be devastating – as bad as any nuclear explosion that this world had seen. The pod certainly had a lot of instruments on it, and some of the tools could be used as attack weapons, if necessary, but he hoped that it wouldn't come to that. *Enough people have died today, both here on this planet and those in a far-off star system.* Within seconds, the planes were in close formation around the pod. One was central, behind the pod and looking straight at it, and the other two were on either side of the pod. A classic battle formation, but still nice to see.

In Arabic, "OK, to whoever is in that flying machine. I have a few things to say. Why can we not see you? You do not show any cockpit. Will you land with us and allow us to escort you to safety?"

"I am showing no cockpit, as you call it, because I do not have one, at least not the glass that you have. I am flying by my instruments. At this time, I am in my commander's chair facing backwards in the direction of my travel. I have a set of screens that allow me to see all around me – I believe that you call it a 360-degree view? My craft is a large size, but it is a survival pod, not a spaceship as such."

Then an Israeli voice cut in, in Hebrew. "This is Eilat airport. You are clear to land, preferably at the far end of the runway, away from the sea areas, if you can."

"Eilat airport, thank you. I am currently being escorted by three fighter planes from Jordan. Can you advise their leadership that I

31

will be your guest and not theirs? This is nothing personal, Jordan. I have my reasons for landing in Israel."

"I know about your escort, unidentified pilot. If you check your radar, you will find that those three will be moving away very soon." As that was said, five Israeli fighters – faster and more deadly-looking – screamed past them all and turned in a tighter curve than the Jordan pilots had just done. Seconds later, they were seated behind the Jordanian planes. "This is the Israeli leader to the Jordanian leader. We will take it from here, thank you."

There was a short burst of chatter between the ground and the Jordanian planes, and then all three Jordanian planes flipped their wings to one side and dropped down, away from the pod and the Israeli flight. The Jordan leader left one passing comment: "Nice flying, by the way, Ben. Shari sends her love. I'll try and catch up with you at the weekend." He laughed and then cut the signal.

The Admiral was curious. "Greetings to the Israeli flight leader. Does the other pilot know you?"

All business-like, the Israeli leader replied with a clear note of tension in his voice, "Yes, he is my brother-in-law. Now your turn. Who are you?"

"It will be better when we are on the ground. For the moment, let me say that I have come to ask Israel a big favour and to look after me. In return, I assure you that I can make myself available to your politicians and your scientific people. In view of the recent catastrophes that this planet has fallen victim to, I urgently need to talk particularly to your science people. They will need some medicines and designs that I can offer to help the population with illnesses and such like, and I need to be able to cleanse the air. Time is of the essence here, so I hope you will not delay me."

"That makes no sense, and right now we have a lot of things going on. We are all very on edge. In case you missed it, we have had a large meteor or something land somewhere around Tibet, and there are a lot of damage reports still coming in. We are not in a playful mood."

"Fair enough. Can you at least listen to me and believe me when I say that it was not a meteor that landed? I will tell you all about it when we get down. Can this be a secure location? I don't want to have your newspapers and other media putting these details about."

"Looking at the shape of your craft, it is already going to be a secure meeting. Why do you say that it was not a meteor that came down? It is reported to be very big and has killed a lot of people." He voiced his concern out of surprise more than anything. *This craft has no windows! What else did it hide?*

"I know, I know. I have been listening to your news items for the last few hours. I know that it was not a meteor because I know exactly what it was that cut through your atmosphere and killed so many."

"And what was it, then?"

"My ship!"

THE STIRRING

IN THE WATER MOVES OUT

The huge shape had left the trench. Its terms of reference were completely different from the life forms that it could see on the surface. It moved very slowly because it did not feel that it needed to go any faster. What was the point? It would still get to the destination, and it had no concept of urgency – if it arrived tomorrow or next year, what did it matter, so long as it arrived?

It found that the oceans were teeming with millions of varieties of life. He looked at each one that he found and was excited. What had he been missing during all of these years? He looked at the life forms on the surface and found just as many varieties. Millions of them, from the very small to the (compared to him) not so small. He investigated everything, above and below the water line, and realised that a lot had changed since he was last looking around.

He stopped moving and wondered where he should be going. He could go to where all of this noise started, but that was a long way away, across all of this ocean that the surface people called the Pacific. He decided that he would turn around and go to where the noise had ended, as it was much closer.

He turned and headed back towards the Philippines and Malaysia. He had an idea that he would check out these places on the way to see what the life forms on the surface had made of them. His experience of the life forms that he had stumbled across is that they usually mess up – they have a sense – a 'view' of the world that is simply at odds with reality. This would be an interesting few days, and then perhaps he could get back to sleep.

Until he could get back to sleep, he connected to the Internet and logged on to millions of computers, finding out what was going on in this world. The more he explored, the more disappointed he was. They have a fantastic level of technology, and yet socially they don't seem to have moved out of a primitive state! They are doomed if they continue with this madness.

THE ADMIRAL

GETS TO MEET SOME OF THE LOCALS

The silence from the accompanying aeroplane was enough to tell the Admiral that he needed to be very careful from here. *I have just given them some monstrous information, and they need some time to work it through.*

After they had approached Tabuk, the other pilot came on the air. "I need you to slow your speed and follow us. We are moving west to go to the Red Sea. From there, we will turn north and move up to Eilat. We are going to try to avoid any public areas, so we are going to go around Eilat and try to land a little way past the town. Your arrival is something that I think we need to keep quiet about until we have established exactly what we are going to be able to do with you and what we tell everyone else in the rest of the world. Do you understand all of that?"

There was a slight but definite change in his voice. The Admiral could sense that this Israeli pilot was now scared and nervous but still professional.

"I understand it very well, pilot. I will land as you have instructed." He waited a few seconds and then added, "Thank you for your assistance."

The other pilot acknowledged his 'thank you' with a gruff "You're welcome. How much can you manoeuvre your craft? We will be going in slow and high, and then I want to come down quite rapidly once we have negotiated around Eilat. Are you going to be comfortable with close formation flying? I would like three of my planes to be tight together, and then you are to be above them, as close as you can get. That will mask you from the ground."

"I have a high degree of control. I can stop where I am and go up, down, left, or right. I was thinking that your craft would be the more limited of us. Let us try. Pick three of your planes and get them into your tight formation: one in front and two at the back. I will then move to be above the rear planes."

The squadron leader gave a few short orders, and then the admiral saw three planes move into the required positions. He stayed at the rear, watching how this was going to work.

They are very good, these pilots, and I suppose that they can call that close, perhaps. Now let's see how they deal with what I call 'close'.

He flew from his position into the area between the two trailing planes. Since they were a few feet apart, he felt that he could position himself directly above the two wings that were closest to each other.

The squadron leader looked, in some shock, as the strange pod hovered above the wings, just inches from them, and matched their speed exactly. Only then did he realise that the pod was big enough to mask all three of the planes.

"I think that's close enough, thank you. You have proved your point. Can you lift yourself away slightly to allow them the chance to move away quickly if they need to?"

"I can, but I thought the point of this was to keep me out of sight? If the lead plane were to drop some twenty feet, and these two planes caught up slightly, then the triangle that they make will make it very difficult to see anything above them, certainly from the ground. I would then move slightly so that I am in the exact centre of their triangle."

Benyamin wondered quite what he was getting into here when their surprise visitor was telling them what to do, and then he worked out that he didn't have a better option.

"Flight 3, take the last message as if given from me, and Flights 4 and 5, move into positions now, please!"

They all changed their positions, and the Admiral moved into the centre of their formation. *So far, so good*, he thought.

They all turned west and flew on until they hit the Gulf of Aqaba. By now, his pod's mapping had caught up with the technology of this planet, and the various place names flashed up on the screens and then were gone. They had slowed now to one hundred miles an hour, still operating at a two-mile altitude. He had to wonder how the planes around him stayed up. The technology was essentially primitive compared to that which he was used to, and he knew that the planes had to maintain a minimum speed or literally fall out of the sky.

As they came up the Red Sea and approached Eilat, he followed the lead planes as they banked left, moving west before they got to Taba, skimming into a little bit of Egypt. They then began a circular route to take them around the western part of Eilat, hopefully at a distance and altitude that no one would be able to take film of them. The formation kept a standard speed, and it wasn't until they were north of Eilat that they started to slow. He slowed down to match their speed and wondered how they stayed in the air. They were doing less than one hundred miles an hour. Only then did he notice the rotating air vents on either side. These planes can hover, he thought! He smiled. What a brilliant idea. "To the gentlemen in the aeroplanes around me, that is some impressive flying. I am delighted to have met you, and once we are done, perhaps I will get to meet you in person."

"Quite possibly. What we need to do is go a little past Eilat and land at a pre-arranged place. We will be met by people. There will be escort vehicles waiting. You just need to follow us."

"As you wish."

Their flight continued for only a few moments, and then the accompanying flight commander spoke in his ear, "You will see a few escort vehicles. Are you able to land close to the large flatbed trailer? We wish to load your craft onto it and then go on our journey."

"Well, the degree of control that my craft has will let me land directly on it, if you will allow me, but what is this with a continuing journey? I thought we were all landing here. I thought I would be meeting people here?"

"This is not a secure location. We plan to go to Israel to one of our most secure locations."

The Admiral checked the pod system and could see that he had fuel left for another few hours. "Pilot, I wish you had told me that earlier. I have enough fuel to fly on if you would be allowed to accompany me. We would be wherever you want within Israel in minutes, not hours." He had picked up the time frames that this planet used quite easily.

"Hold on." There was a lot of air-to-ground chatter, and the Admiral presumed that the ground crew had to ask other ground-based personnel either for permission or for their ideas on the possible change in plan.

A new voice came on his radio. "Pilot of the unidentified craft. My name is Ori, for the moment that will have to do. I understand that your name is Avraham? If that is so, then I would like a very brief summary of who you are, how you got here, and why you have chosen to land in Israel."

The Admiral wondered who this 'Ori' person was, but decided to answer as best he could. "Pilot Ori, I will not start to meet you or your people by telling lies. I have lost much today, as has the population of this planet. If you should choose not to believe me, then it is a loss for both of us."

"My name, in my native language, is close to Avraham Avraham. I am an officer – an admiral in the space navy of my world – from a star system some distance from where you are that has used what you call jump gates for the last few centuries. We have travelled far, but even we cannot explore all of the space around us in a few days. We have our own problems with aliens and pirates, as you seem to do on this world. I have just escaped from a fatal battle in one of our star systems. What you saw crashing down to your planet was not some rogue asteroid but the remains of my ship. In your measurements, it was three miles long and about half a mile wide. It was never designed for travelling through any planet's atmosphere, which is why it fell apart so readily and so messily. It was already half destroyed by the time I came through the jump gate, but going through your atmosphere completely wrecked it. I apologise for all that have been killed or injured – that was not my plan, but the ship was completely out of my control when we came out of the gate."

"I landed in the area close to where the ship came down. I stayed out of sight for a few hours while I monitored your news and information channels. I have never been to this land before, and I doubt that any of my people have, so I didn't know anything about it. I watched and heard all of the details of the ship coming down and that growing casualty list in many languages. Try to imagine my surprise when I heard this Hebrew, or as you call it, Ivrit. It is as close to the language used on my Homeworld as I can make it. I think it is only the vowels that make it slightly different. All of these other languages are wildly different. I felt that I needed to get to Israel, to make enquiries here, and see if I can somehow make an attempt to be useful, to make a life, perhaps a living here. There might well be other survivors along the path of the ship, but I have not heard anything clear from anyone yet. It may be that, like me, they will go quiet for a few days until they know what is happening." How to finish? "At this time, I have nothing to offer you in terms of assurances. You know as much about me as I do about you. I only hope that this will change."

There was silence at the other end, presumably while some conferencing was going on. It took a few minutes; all the while, he and the jets were hovering over the proposed landing spot, some way north of Eilat airport.

"To the leader of the accompanying planes, how is your fuel situation? I calculate that you have used a lot while chasing me. I am good for another few hours."

"Avraham, I have a name. Benyamin Ori will do for now. Our fuel is getting low, but we will be good for another short trip within Israel." He finished the message with a quiet "I hope" that only he heard. "Pilot Avraham, this is Ori. You must understand that we are nervous about your arrival. If you have checked out the news around Israel, you will perhaps see that a lot of people spend a lot of time trying to do us harm. We are wary of all unannounced visits, wherever they are from. If I were to allow you to proceed into Israel,

41

then you must be aware that we will be treating you with suspicion until you can land and we can have this conversation directly, Man to Man."

"Benyamin Ori, you worry too much. I accept all that you say. If these fighter planes are to accompany me, then so be it. It is not my intention to harm anyone here at all. My ship has caused too much damage already. I need to see if I can help with any rescue plans that may be put in place, although I find it difficult to see what I can do, given what I believe is the current level of technology on this planet."

THE ADMIRAL

A STRANGER IN A STRANGE LAND

Silence for a few moments and then "Avraham, we are cleared to proceed north. We will climb to five thousand feet and head to a place called Dimona. There is a small airfield there. Once we get there, follow me in. We will not be using the runway. To the south of the runway, there are blocks of buildings, and we will land very close to them. I will confirm it once we get closer, but we have a large building, which we call an aircraft hangar. I want to see if you can fly into that and then park your vehicle. It will get you and your craft out of sight of prying eyes."

"Benyamin Ori, I am ready to go. Lead on, and I will follow."

The Admiral then saw the lead two fighters turn and move away, heading directly north. Since there was no reason to follow any roads, they just flew in a straight line, heading towards this mysterious town, Dimona. "Avraham, would you follow, please? We will then follow you."

"As you wish." He touched the controls, and the pod rose up and moved to fall in behind the fighters. He could see the others fall in behind him. *OK*, he thought. *You still don't trust me, do you?* He chuckled. *Fair enough. I wouldn't trust me either.*

He smiled. *So far, so good. Now just don't mess it up, somehow.*

The formation flew up into the sky. Once they reached their altitude, the leader of the fighters came on with "We will now start our descent. We needed to be quite high, to make sure nobody saw us, but it's only a very short journey."

"As you wish." He didn't know where he was going, but just followed the fighters in.

"We will come in low for the last part. The southern part of the airport is shut down for our arrival. This will hopefully mean that people will not see us turn up. We want this kept quiet, not all over the Internet."

They dropped their altitude and speed, coming in from the southeast, across what looked to the Admiral to be desert. It wasn't until much later that Avraham was told they had not been given permission to fly directly over the nuclear research facilities just outside of Dimona. They had needed to make a small course correction to go around that facility.

Since the buildings were south of the runway, it meant that the formation could come in slow and quietly, landing directly on the ground around the buildings. *These pilots are good,* he thought.

There were three of these hangars next to each other. Only the middle one had the large doors open. "Avraham, can you see the large hangar, the one with the doors open?"

"I see it. Do you want me to go straight in?"

"Yes, please, nice and easy."

He then noticed that two of the fighters were not on the ground, but were in fact still orbiting the airport. *That instruction has been given on a frequency that I was not following.* They were staying away from the north part and were running a tandem route, which meant that there was always at least one fighter close to the airport. He smiled and added *These pilots are very good!*

With very little worry, he turned the pod around and backed through the doors and into the hangar. It was almost empty, and he

44

set it down in the middle of the floor. It was very big, this 'hangar' as they called it. He knew that the other pods could be accommodated in this and the other two. He powered down most of the systems, intending to save what fuel remained. Once he had done that, he relaxed. For the first time since he had arrived on this planet, he hoped that he was safe.

A voice cut into his thoughts. "Avraham, I am outside your vehicle. Do you want to meet now?"

There was a slight tension in the voice, but the Admiral hoped that it was because they were all tired. They have just had the experience of a lifetime!

"I will be right there." He checked one more time that the systems were correctly operating, and then opened the door to the pod. He stretched and slowly climbed out. The air was warm, and he noticed that the large doors were closing. *Well, here we go.*

He walked down the ramp towards the small group of people. As he got closer, he saw that they were just like him in form and in size – two legs, two arms, a body, and if the last few hours were a good example, then a very capable brain!

One of the men stepped forward. "Avraham, I am Benyamin Ori. Welcome to Israel." He was smiling and offering one of his arms out, but Avraham could sense enormous tension in the voice. Ah, yes, this custom of shaking hands. The Admiral put his arm out and they held each other's hands. Nothing bad happened, so the Admiral decided to go a little further. He dropped his arm and gave Benyamin a gentle hug.

He was wondering whether he should give him a kiss, but decided that it might be better to wait until he had seen what these people accepted in their social lives. On Homeworld, people gave each other kisses, whether they were male or female. It was just something that was done, rather like this shaking of hands.

Benyamin was clearly surprised by the hug but didn't fight back, so the Admiral hoped he hadn't overstepped any social mark.

"Thank you." He looked at the other pilots and spotted that three were women. *Women pilots? Now that's a first!* "Women pilots! That is an excellent thing for me to find. Thank you all for looking after me." While he continued talking, he moved among them, shaking hands with everyone he met, at the same time being very careful that he didn't hurt anyone. "I cannot tell you how nervous I was about coming in. What now? I assume that you have people I should meet, talk to, and tell my life story over and over again – I will get bored with doing that before very long. Are the other pilots still circling the airport?"

"They are, and they will stay there for a while. Security at the moment is at its highest level. The people who want to talk to you won't be here for a while, so until that time, we are under instructions to try to make you comfortable, but you will have to stay here for a little while. The facilities here are basic, but you could have a wash if you like. We have showers here."

"The facilities in my pod are even more basic, I'm sure. I would love to have a clean-up, a shower. Do you think someone could find some other clothes for me – some clean clothes? I have been living in these for the last week!"

"I'll see what we can do. Come with me, while we get out of the hangar and into a more relaxing area. You will have to tell me what you eat or drink – it is likely to be different here from what you are used to."

They walked away from the pod, and he wondered when he might see it again. They went through some other doors and walked down a short corridor. Opening another door, they entered a small room with chairs and tables. There were some machines against one of the walls, and the Admiral assumed that they were the drink dispensers, where people got this 'tea' and 'coffee'.

"We are in the social part of this area of the airport. This part has nothing to do with the public traffic, so you won't be seeing any tourists passing around. We have locked down this whole area, so at the moment there is only you and me here, with the airport security under instructions to close it down – no one in or out unless authorised. We come here to relax, catch up on news, get some drinks, something to eat, that sort of thing. Do you have any idea what it is that you can eat or drink here? I'd hate to think that you have come here and the first thing we do is try to poison you."

"From what I know, you have water. It is a basic liquid, and it appears everywhere that I have visited. It has to be present for all things to grow, to survive. Whatever else you have, this 'tea', 'coffee', I will just have to try. I think my body is resistant to a lot of toxins, so I will have a little of everything, I suppose, and tell you what I don't like. Nothing complicated – I am sure whatever will kill you will at least make me very ill. Can I start with water, please?"

"Let me show you how this works." They both walked over to the machines, and he got a plastic cup and held it under the nozzle, and pressed the lever for the water to come out. "Any time, just help yourself. Here, you take this, and I'll get myself another one." He handed Avraham the cup and filled another. Avraham downed the contents and then filled up and drained three more cups. He hadn't realised how thirsty he was. They both walked to some seats and sat down. Benyamin was wondering what to say next. "We have an expression. When we meet people, people that we may know or people that we may not know, we toast each other. We usually say 'Cheers!' which translates to 'Good health.' I suppose. So, pilot of the strange new craft, Avraham, from the people of Earth to you, I say 'Cheers!'." He stopped and looked at this stranger – could he trust him, this alien, who said he was from another star system? "Whatever we may be, we are not fools, you and I. I have to assume that you are some sort of soldier, as are we all in Israel. We have had to be. I don't know how your craft works, but I'm sure we'll eventually get you to describe the principles involved. We probably don't have the level of technology yet, and may not for many years. We also know nothing of your Home world, and you know little of our world, this Earth, or the complexities of our different societies."

"All of that is true, Benyamin. I must admit that not knowing much at all about your social structures, your internal struggles..." he trailed off. "I do not like to be in the dark about things. The more I know then the more I can understand and react to."

At that time, there was a lot of screeching of brakes outside, and in a few seconds, a squad of heavily armed soldiers had stormed in, isolating the entire area. No one in and no one out. "Ah, it looks as if the first of our visitors has arrived. Avraham, try not to panic, but there will be a lot of new people turning up now. Some are our science people, and then some are our political people." He felt a tension creep into his voice. Quietly, he volunteered, "It's the political ones you need to pay attention to. They are the ones who control everything that is happening. The soldiers, including me, do as we are instructed by the politicians."

"Benyamin, my friend, if I may call you that, thank you for your advice. Yes, I am a professional soldier, and have been for most of my life. It has been an interesting life so far, and one I intend to continue with, so I am not about to do something stupid to upset those whom I need to be careful around. That said, I would appreciate any advice that you may be able to give. I may well say something 'simple' that upsets someone else, for the most ridiculous of reasons."

"True enough. I would be honoured for you and me to call each other friends, but in our society, we usually earn friendship. It is not something that we entertain lightly. Oh, we have levels of friendships, where you might meet someone briefly and never see them again. Other times, you may meet someone for the first time, and it is as if you have known them all of your life. That is rare, but does happen." He smiled at this stranger, this visitor. "Let us - you and I - see how our friendship develops, shall we?"

THE ADMIRAL

MEETS THE FIRST GROUP

With the area secured, a group of civilians then marched into the room. "Right, let's get this started, shall we. Who is in charge here?"

Benyamin stood up. "That would be me, then, sir. Aluf mishne Benyamin Ori."

The newcomer was surprised. "And what pray is an Aluf mishne doing looking after this mission?"

He shrugged. "It is not complicated. We were on a recon mission, and the call came in. We were the nearest flight that was armed and ready, fuelled up, so we got the first call. Now, sir, if you would introduce yourself?"

"Sorry. I am Deputy Director Orstovsky of Kidon."

Benyamin exploded. "What? Kidon? What the hell are you doing here?"

He shrugged. "Hey! Relax, please. Like you, we were the closest to the area. It would normally have been Mossad – they are on their way, of course – but we needed to get this area fenced off quickly, and we were asked to attend. There are fifteen of us, and we will be locking down this whole area."

"This is outrageous." He hissed. "How the hell am I supposed to explain to our guest that you are our guards, our prisoners, and if necessary, our executioners? I know of Kidon, so be careful how you answer, Mister Deputy Director."

50

With a hint of menace in his voice, he said, "General, stop it. We have our instructions, as do you. At the top of our list is to secure the area. Once that has been done, then we move on to Phase Two. There is a lot of shit flying around at the moment, so please don't make things more difficult. Now, will you please introduce me to our guest?"

Reluctantly, Benyamin turned to Avraham. "Avraham, allow me to introduce one of the many people that I am sure you will be meeting over the next few days. This is Kidon Deputy Director Orstovsky."

"Kidon? That name I am not familiar with. I assume that Israel, like many other countries, has many different departments in its structure. I am Admiral Avraham Avraham – that is the closest that my title and my name will translate into your language."

"Admiral Avraham, I am sure that this will not be the last time you hear it, but welcome to the planet known as Earth. You must surely realise that we, the population here, have many questions for you, and I am sure that you have many questions for us as well. I will add that until we have established that what you say is true, then we will consider you with great interest, and possibly some suspicion. The planet has recently had a lot of its population destroyed, and here I am in front of the person who admits that it is their ship that did it!"

"Deputy Director, I will answer what I can of your questions. I hope in return that you and your people will answer some of mine. That will do for a start."

"I'm confused. What do you mean by 'for a start'?"

He shrugged. Was this person going to make life difficult? "I mean what I say. You and I have many questions still to ask. How many different sexes does your species have? Do you allow women to own businesses, or are they to be regarded as someone less than a man? Do you have many established religions, and which ones are the main ones, and where are they based? Are there local ones as well as national or international ones? Do you permit men and women to enjoy each other's company as and when they like, or is there some established system that says men and women may only be together if they have formally recognised their relationship? Do you accept what I might call same-sex relationships, or is that another thing that is frowned upon? I could probably go on, but that will do. I want to know about your world as you will want to know about mine."

Orstovsky actually coughed. Benyamin and Avraham exchanged glances, smiled, and both shrugged. Benyamin stepped in with "We were just about to try to arrange some food. My crew hasn't relaxed for over a day, or eaten almost as long. Given what our guest has admitted, I doubt that he has tasted Earth food before, and certainly not Israeli food! We were just about to send out for some local food – we will obviously get it arranged through the town, and we will meet the delivery at the airport. We will keep it quiet, and I think your team should stay clandestine for a while."

"Clandestine? I'm not sure I understand the meaning of that term, in the current usage."

"It means that the Kidon teams are high-profile – they stand out. They are easily noticed. If we are to do this properly, then we need to be invisible to everyone."

"Ah, I see. I understand. Perhaps soon we can talk about whether you can start to use some of my technology? It may take a while to get you familiar with some aspects of it, but since I believe that I am to be on this planet for some time, I will need to make my services available to some people. Given the scale of the damage caused by my ship coming down, I must tell you now that some of my technology has been designed to clean the air – it uses methods that you are not yet familiar with but should be able to pick up quickly enough. I am keen to start on this before the entire planet gets covered by the fallout."

Orstovsky was no fool. He was getting constant updates and knew that the planet had a week at most before the air became too poisonous to breathe. "Admiral, once our science people arrive, I would like you to outline what you can do on that front. I am not aware of too many specifics, but the fallout is spreading, and the latest estimates are that we have a few days, perhaps a week, before we are all wiped out."

Benyamin turned to him, "A week!" He turned away and took a large drink of water. "Shit!" he said quietly.

"Mossad will be another hour, so I suggest that we make ourselves comfortable."

"An hour? Where the hell are they coming from?" Benyamin was feeling flat after all of the recent excitement. "Can we get a secure line to the Prime Minister?"

"We are trying to arrange that. Given the current situation, the entire Cabinet has gone into a War footing until we get a clear picture of what is happening. Because of that, it is now difficult to contact him."

"Well, that's just great. The one person that we need to speak to, urgently, and he is not available? Brilliant!"

Avraham asked, "Can I get that clean up? A shower, perhaps, and some clean clothes?"

Benyamin volunteered. "That's a great idea. Since nobody wants to talk to us, we might as well go and get cleaned up! Orstovsky, you may send guards, but our visitor and my team could do with a clean-up. If you are going to be here, can you send out for some food? I don't care what, but enough for all of us." He turned to his pilots. "People, time for a shower!"

THE ADMIRAL

HAS A SHOWER, WITH NEW FRIENDS

"Hooray for that. I was beginning to wonder if I'd ever see water again!" said one of the ladies. They all picked themselves up and moved off, following Benyamin, who seemed to know his way around the place. Five of the security team followed them. They walked through a couple of doors and entered the shower rooms. Benyamin turned to the guards. "Unless you are joining us, I'd advise that you stay outside. It's likely to get very wet in there. I know this place very well, and this is the only door in and out. None of us will be going anywhere." He then turned to his guest. "Avraham, I need to ask. Are you particular about how you have a shower? The people in my flight have all seen each other in training sessions and the clean-up showers afterwards. We tend not to bother whether we shower alone or with other men, or with other women. Is that acceptable to you?"

He smiled at Benyamin. "This goes back to what I was saying to your Deputy Director. I have a lot to talk about with all of you, and I would ask that you talk to me, to educate me about this planet, what goes on, who is important, and why, that sort of thing. There is a lot to cover. In my world, we shower with whoever we like, and we have same-sex relationships or mixed-sex relationships. We take it all in our stride."

"Fantastic." He turned to the others. "People, I would like to propose that our team has a new member. Let's all go and get wet."

There was a 'Whoo-hoop!' from the team. They went into the changing rooms and they all immediately started to strip. It was not at all sexual in content, and they all were disrobed within 30 seconds, and they all picked up towels and walked into the showers. Men and women, all completely naked and at home with each other's company.

They all stood under the showers for some time, washing off the dirt of the last few days. Like a meal, these people didn't know when they would have their next shower.

One of the women approached Avraham and asked, "I am called Golda. May I call you Avraham?"

He smiled warmly. For an alien species, these women were very attractive. "Of course you may. I would be delighted if you did, if I may call you Golda?"

"Yes, you may." She smiled. "If you are not human, then we will have to work out what you are. To us, you are an alien, even if you look remarkably like any other person from Earth. To you, it is we who are the aliens. We must try to explore each other and find our common areas, and since you look like us, I have to wonder if there is perhaps some common ancestry, even if it was millions of years ago."

"I agree. That is a thought that has been going through my mind. I welcome any of your questions."

She continued, "Do you have an age? We have the concept that this planet takes what we call a year to go around the Sun, our star. We usually age people in years. Are you able to translate our years and your years, so we can understand how old you are, in our terms?"

"I have looked into this. With your projected timelines, I am about three hundred of your years old." He looked at the woman. "How old are you?"

She gasped. "I'm twenty-eight!" and then she laughed.

Benyamin looked at this stranger and actually coughed. *Three hundred? He doesn't look much older than me!* "Three–hundred? I think I need a strong drink. On this planet, we know of some people who make it to one hundred years, perhaps a little more, but not many. Most of the people die in their sixties to their eighties, and there are a lot of people who die very young – childhood diseases, things that we simply don't have a cure for yet."

"Ah, then that is a wonderful opportunity for me. I can use some of my experience and hopefully talk with your chemists and researchers. We eliminated over ninety percent of our ailments a few centuries ago."

Out loud, Golda turned to them all and smiled, a mischievous smile. "When we get the sleeping arrangements sorted out, I want him first!"

All of the others turned and smiled. Benyamin laughed, "I was wondering how long it would take you."

A slightly confused, but still relaxed Avraham said, "I don't think that I followed that. Did something happen that I don't know about? What is this 'want me first'?"

Benyamin moved to Avraham and put his arm around his shoulder. "Avraham, my friend, the people in my squadron all know each other very well. *Very* well. We have few secrets. Yes, we have all shared each other's bed at times. Golda here thinks that she would like to get to know you very well in bed. Although you and I don't understand everything about each other yet, please take this as a compliment. She likes you. As I think, do we all."

Avraham noticed immediately the change in the tone of the voice. He had just been called 'friend' by Benyamin.

"Benyamin, I am honoured to be so accepted by your people, your friends. Truly, I like you all, and I hope that over time we can all get to know each other better." He turned to Benyamin and put on his serious face. "There is something that I must tell you, and you need to hear it, now. You need to understand some things about me. I come from a very different world to yours. I have said that I am a lot older than all of you, and that is true. I am also different, physically, to all of you."

Without knowing why, they all took a step back. Golda asked, a hint of ice showing in her voice, "You'll have to explain that, and clearly. We all trust each other, here, with our very lives. We cannot have trust in someone who hides things or lies to us."

"I have not lied to you. I have no plans to lie to you. I have not yet told you everything about me, that is all. I have a lot to tell. Think about it. How do I squeeze the last few hundred years of my existence into the last few decades of yours?" He paused, hoping that this next bit would go better than he was expecting it to. "I am much stronger than any of you. When you and I shook hands, Benyamin, I had to really concentrate so that I didn't crush your hand into pulp. I have not yet had a case where I can work out how much stronger I am than any of you, but I can easily do some serious damage to you, without realising it." He turned to Golda and, being very careful, gave her an embrace. "With that in mind, I do not know if you and I can be in bed with each other, as I may seriously damage you."

She gasped. They all stopped and looked at him. For a few seconds, no one moved and no one talked. Golda then laughed and decided to try again anyway. "What happens if you don't actually move? What happens if I am on top?"

He looked at her, more than a little surprised. "I don't know. In my world, it is usual for the Man –"

"No, no, no. Stop. This isn't your world. It's ours. Hold still. I want to try something." She went to him and held one of his wrists, putting a thumb against his forearm. She lifted the thumb and saw no change. "Ben, we may have a problem. There is no change to the blood flow. I would have expected the skin covering to go white while I was applying pressure, and then recover when I took my thumb off. Nothing changed. This means that he is immensely strong. His skin covering is of the order of at least ten times stronger than ours, if not more."

"That means that we have to be careful. Let's keep this between us, people, for now. I'm pretty sure that our Assistant Director will not like that sort of news."

"What does all of this mean? I'm confused."

"Don't be. It means that Golda will still want to see what happens in bed; it's just that she will have to do most of the work, since you have just admitted that you could easily damage her. We will be watching very closely, and if any of us thinks that you are getting out of hand, then we will have to step in. I do not want any of my team to be hurt."

"You will be watching? While Golda and I are engaging in close sex?"

"Oh – is that a problem? We really have few secrets from each other, and yes, we have all been engaged in sex in the same room with each other. For us, it is acceptable. For some people, it is not, but in our case, we enjoy the moment."

Avraham was looking at them all and felt that here, with these people, he was truly safe. He knew that he trusted them and felt that they were beginning to trust him. "Benyamin, it is not a problem. I mention it only to make sure that you and I understand each other. Does this mean that the other ladies will be present as well?"

One of them voiced, "We wouldn't miss it, Avvy. Wouldn't miss it for the world!" She then thought of what she had said. "Oh, I'm sorry – perhaps you don't like your name being shortened. In this world, we do it all the time. Benyamin here is often called Ben, and Avraham just becomes 'Avvy'. Are you OK with that?"

"I am very happy with that. We maintain our full names for anything formal, when we are on duty, that sort of thing, but outside of that, we often call each other by shortened names. I have been called Avv for many decades, and Avvy is a nice change."

Benyamin knew that they had spent enough time on this. "People, we are done here. Dry and dress. I'm pretty sure our next visitors will be close to here by now. We have some clean clothes for all of us. These sites usually have spare kit for just this kind of thing."

"Excellent. Ben, you and I have only just met each other, but I must say that I am impressed with your professionalism." He turned to the rest of the team. "And that goes for the rest of you as well. I am a professional soldier, and I know what to see in another professional soldier. I know what to look for, what to expect. You have all shown all of the qualities that I would wish for. For that I am grateful." He paused. How could he say this next part so that he did not alienate all of these good people? "I must ask that you now think on what has happened, and what must happen over the next few days. Although I did not know it when I appeared between your Earth and your Moon, my arrival has doomed your planet – well, your race – I am sure that your planet will continue, whether Man is there or not - to extinction, unless I am allowed to perform certain actions within the next few days. I do not say this lightly. I am most sincere in what I say. Your planet is doomed unless I am given a free hand to do things that you most likely will not understand. To me, this is not an issue, but I feel that it may be for some of you. This is your call. I would ask that, whatever questions you may have, you ask me, now, today. Do not hold something back, because to do so may jeopardise the safety of your species." He paused again to look around the locker room, with these people all naked, with no secrets. *They need to understand – They MUST understand!*

The enormity of what had been said hit home to Benyamin hard. He realised that he was in the presence of Aladdin's genie, the Goose that laid the Golden Egg, and more – much more. If even half of this was true, then Israel was at the same time under threat from its neighbours even while they held the answer to – at a best guess – half of the troubles of the scientific world, including a way to stop the latest catastrophe from getting any worse. It was too late to undo any of the damage, but if there was a possibility of saving what was left of Humanity, then it had to be taken.

One of the other women stepped forward, wondering if what she was about to say would go for or against them. "Avraham, I am Tsipi. I cannot be sure, but I hope that I will speak for all here. None of us here is a fool – we are considered to be among the top intelligent people in this country, possibly, given the current circumstances, amongst the world. We realise that your arrival is chance – you could have arrived at one of another thousand worlds, but you came here, by sheer chance. We, from this planet, must deal with this. Additionally, your arrival has not been smooth. By your own admission, your arrival may well be the end-all of human life on this planet. We cannot take that too lightly. We must admit that we on this planet simply do not have the technology to make the issues that you have caused go away. If you are telling us how we can correct the problem, then we must listen. We must have, as our goal, the survival of the species. That is what we are interested in."

She turned to the others and received nods from each of them. Benyamin signalled that she should continue.

"Avraham, we may well find out that some of our political people cannot see too far – they see only into the next hour or the next day, they cannot appreciate that they need to be seeing into next year or possibly further. That is a fault of our politicians, not of us. I hope that I speak for all of us here, in this room, when I say that we are with you. If we have to go against our politicians, then so be it. The end of the game is that we survive. We are not interested in playing political niceties."

"Tsipi, I am delighted to have made your acquaintance. If what you are saying is true, then I will ask a lot of you, probably sooner than you would think. The people that I have to meet will only get in the way. They will delay my activities, which will jeopardise the future for all of you." He made a decision. "I need to get to my pod, and I need to contact all of my people, any that have survived, that is. I need as many pods as I can to do a cleaning of the air." He looked around. "By myself, I will fail. Think on that."

"Do you need to get to the pod now? I'd prefer it if we didn't wait."

"Benyamin, there is one more thing that you should know. I am in communication with my pod. In addition to being very strong, I have had a lot of surgery done on me. This was done over many years. I have been artificially enhanced in many areas, one of those areas being adding to my strength, and another is creating a sophisticated communication system within my body. I need to activate some of the pod systems. It will need to scan for any more that have arrived. You should also know that some scramjets came down at the same time that I did, but I do not know where they have landed. They can help with the cleaning of the air, if I can find them. These are fighters. As good as you consider your fighters to be, take my word for it when I say that these are better. A lot better."

Benyamin wondered what they were getting into, but said, "Fine, do it. The earlier you make a start, the quicker we can get this set of repairs underway."

"I don't need to voice anything. It works off impulses from my brain." Only a few seconds of silence passed, and then "It is done. The pod is searching, and will tell me when it has located the others."

"Then we had better get finished here and move back to the common room. By now, half of Israel will be wondering where we have got to!"

THE ADMIRAL

MEETS THE SECOND GROUP

They moved out from the showering area and retraced their steps back to the common room. When they entered, there were some new faces, and Orstovsky stepped forward. "So nice of you to join us. We have Mossad with us, and they would like to interview you and our guest!" He was clearly annoyed that they had taken themselves off, and Benyamin wondered if the Mossad people had told him off for letting them go. *Not my circus*, he thought.

He noticed the table was laid out with the food. Had they been in the showers that long? "Is that for us?" He asked the Deputy Director. He nodded. "Good. Tuck in, people, we don't know when the next meal will be."

Someone new stepped in, "Major Ori? My name is Kaplinski, of Mossad. I have had a brief update from the Deputy Director here. We have our instructions, which are to take all of you to Jerusalem. The Prime Minister is waiting."

"I thought the Prime Minister was making himself hard to contact?"

He smiled. "Only for some people, Major. For some other people, we manage to get through, somehow."

Benyamin knew that going to Jerusalem was out of the question. If what Avraham had said was true, then that would waste yet more hours. "We don't have the time to waste getting to Jerusalem. We need the Prime Minister at the end of a phone or a video connection. There are things that must be started now, or we may well all be lost."

"I don't follow. 'All be lost'? What does that mean?"

"Stop it. You are not stupid. This visitor has caused a lot of damage by his arrival, and from what we have discussed, he can also do some things to stop it from getting worse. If we want to save what is left of humanity, then we must start now, not in a day or two, when we have gone through all of the political hoops. The situation is already critical – to delay even a few hours may mean that we cannot repair the damage done in enough time!"

"I have my orders, Major."

"Then get whoever gave you your orders on the phone! They need to be updated, now!"

The Mossad leader went to a small group of his people and had quiet words. He then came back and said, "I don't know what you are planning, Major, but my instructions stand. I need all of you ready to go to Jerusalem." He raised his arm, and suddenly all of the armed guards in the room lifted their weapons and pointed at the group. "I hope you are not going to be awkward in this, Major?"

Avraham spoke to Benyamin. "Ben, I am confused. I thought you had just mentioned that we have to start some urgent action, and yet this person is saying that we all need to go to Jerusalem, and lose the best part of a day in travel and meetings?"

"That's about it. I'm sorry, but the situation is getting out of my control."

Addressing the Mossad leader directly, Avraham said, "You, the Mossad person. Kaplinski? I doubt that you are a fool, but when you are told that time is of the essence and you simply ignore it, then I have to wonder. My name is Admiral Avraham Avraham, and I probably outrank you, but we cannot lose time in attempting to stop the scale of the disaster that I have brought on you. You will let me get to my pod and begin to put things right, and I must start now."

"No, I don't think that's going to happen. Your pod is now under guard, sealed off."

He sighed. These people were going to make it a *very* difficult day. "Ben, do you remember that I mentioned that I had been surgically enhanced?"

Wondering where this might be leading, all he could say was "Yes."

"One of the changes was strength, but another was speed."

Although nobody actually saw it happen, Avraham then vanished from sight for a few seconds. When he reappeared, all of the guns in the room that were pointed towards them were wrecked, with the barrels twisted out of shape. Even the small side arms were ripped from their holsters and crushed, and every knife had been collected and lay on the floor in front of Kaplinski.

Standing directly in front of the Mossad leader, he said, a menace in his voice, "So, this is the situation. You will get back on the line to whoever you have to talk to and get an updated set of instructions. I can take out all of you if I wanted to, but I don't. That is not why I am here. I *will* go back to my pod, I *will* get into my pod, and I *will* make a lot of things happen. You will not stop me, because any delay from you is just stupid. You will cause the death of the rest of your Humanity! Ben, to try to keep these idiots happy, would you like to come with me? Perhaps Golda, too? I can take up to four people, including me. Tsipi, would you like to join us?"

"Would I? Oh wow, you'd better believe it."

Benyamin said, "I think that it is time for you to get the Prime Minister. Now." He looked hard at Mossad. "Before this situation gets out of hand."

"I'll see what I can do." He walked away, back to his team. They held a short conference, and one of the team members spoke heatedly on the radio.

Kaplinski came back towards Benyamin. "It will be a couple of minutes." He turned to Avraham. "You say that you are an Admiral, where do you come from? While we wait, tell me what you were in charge of."

Sensing a change, a softening in the tone, Avraham relaxed a little. "Certainly. I had a fleet of what we call Galaxy Class Starships. Each of them has a crew of perhaps six thousand people, over one hundred fission engines. The dimensions are the same for each – about three miles long and half a mile wide, so they are big – the biggest that I have ever seen. Other races that I have had contact with have Starships, but nowhere near as big. I was in charge of ten of these and more than a hundred smaller ships, with varying crew numbers. Why do you ask?"

Ignoring the last question, he continued, "So in rough numbers, you were in charge of at least sixty thousand crew, and it might be nearer to one hundred thousand, if you add in all of the extra ships?

"That is possible, yes."

Kaplinski smiled. "Delighted to meet you, Admiral. Before I joined Mossad, I was an Aluf in the Israeli Navy, so I think that makes us equal rank." He stuck out his hand, and Avraham looked at Benyamin in a confused way, who just nodded. Avraham shook hands, taking great care not to hurt him.

THE ADMIRAL

MEETS THE THIRD GROUP

At that point, the doors opened, and some more armed troopers came in, guns at the ready. Kaplinski shouted to them all, "Stand easy. There are no threats in here. Maintain the perimeter." Then a tall civilian walked in behind them. All that Avraham heard was a hiss from his new friends and a 'Shit!' from Benyamin. The newcomer walked to Kaplinski, and they shook hands automatically. They clearly knew each other well.

"Ram, my friend. How are you keeping these days?"

"Fine, Prime Minister, fine. We must catch up sometime?"

"Agreed. Now, to our guest?"

They both turned and faced Avraham. Kaplinski said, "Admiral Avraham, this is Michael Rothberg, the Prime Minister of Israel."

Shaking hands, the two people assessed each other. "Prime Minister, I am delighted to meet you. A little confused, perhaps, but delighted, nevertheless."

"I must apologise for the confusion. Ram here said that he needed to be very sure of what was going on before I entered the building. He may have been a little aggressive, but there was a reason."

"Ah, I see! This was what you call a ruse, to see if you could find out things from me? Ha! Very good. Well done. With that in mind, then I must agree with your plan."

Rothburg turned to the flight team. "Major Ori? Good to see you. Thank you for looking after the day so well. This has been a day that we will all remember, but it could have been a lot worse."

Avraham said, "Worse? How could it have been worse?"

"We could have shot you down. You should consider yourself lucky that this flight team were the ones to find you. I'm sure that most of the military is very jumpy at the moment, and some are likely to shoot first and tidy up the paperwork afterwards."

"Good point." He smiled at Benyamin. "Thank you for not shooting me down."

He shrugged and smiled. "My pleasure."

"Now, to business. I have been getting a lot of real-time information on what is happening around the world, and I have been getting updates from the people here while we were coming in. You say that you have a way to stop the damage from getting worse? Since I know full well that if we do nothing, then we are all lost, then we must do something, whatever it is! Explain what you can do, if you can, and then what do you need from us? I appreciate that the time here is very short. The latest estimates are that we have perhaps three days before this becomes impossible to change, so I am up for any ideas as to how we can stay alive."

"Prime Minister, I am glad to meet a man of action. Immediately, I need access to my pod. I have to call everyone that I can find of my crew, identify where they are, and create a strategy. I am more than happy for these people to accompany me." He indicated Ben, Golda and Tsipi.

"Good." He laughed. "Does anyone recall Star Trek?" He looked around and saw most of the people raise their hands. He turned back to Ben. "General, make it so!"

Ben saluted as best he could, out of uniform. Here was a true leader, one who knew when to lead and when to delegate. "Avvy, can we start?"

"Of course we can. Do I have access to my pod?" The Mossad leader nodded. "Then let's go. This is going to be one very long day."

They all left the room and walked behind Ben and Avvy. Going through the corridors only took a moment, and then they were there, in the hangar, with the pod in front of them. Avraham turned to the Prime Minister. "There is one thing, and I don't know enough to guess if you can achieve it. My fuel is based on your Uranium 235. I use a more refined version of what you have – a much higher grade – but I believe that the U235 is the only available fuel that I can use. I am running low and will most probably not have enough to maintain the cleansing process for more than a few hours. Do you have any idea where I can get about a half a ton of U235? I would need it broken down into smaller containers, perhaps twenty-five pounds in each container?"

He looked at Kaplinski, who looked in turn at the Deputy Director. They all shrugged, with a *What the hell?* look on their faces. He turned back to Avraham. "Let me make some calls. I'll get back to you in an hour."

"Fair enough. Now, to the pod." As they approached it, the entrance slid open, allowing the ramp to descend. The four of them walked up it, three of them wondering how this was going to work out, and one of them wondering if he could manage to pull this off.

As they got to the top, Avraham introduced them to the work area. "There are four seats. That one is mine." He indicated the master chair. "The toilet facilities are basic, but behind the opposite chair and down one level. It is easy to operate. There are limited

70

food and drink supplies, but for the next couple of hours, we will not be moving." He turned back to the others. "Perhaps you could arrange a supply of basic Earth food and drink? When the other pods get here, I will need them stocked as well. I will also need more supplies for the other pods that I hope to find and get brought over here from South America. We have a lot to do..." He trailed off and walked back down to the Prime Minister. "I need to do a great deal in as short a time as possible. I would prefer it if you could assist in some way."

"I will help where I can, of course. You need to let me know what it is that you want, though."

"I will be making a lot of calls and trying to see who has managed to land in one piece with a pod that I can use. If I get them, then I hope that they will have enough fuel to get here. They will be coming in fast and high. They will need to go up to an orbital height and then across to us, and then down to meet us. You must make your defence system aware of them. One of the first things that I need to do is activate a small number of pods that came down with me. They are at the moment on the seabed in the Gulf. I need to get them here and park them in the adjacent hangars, so can somebody see about making some space in them? I can give you the security codes that they will be sending, so that you don't shoot at them, but the fewer people who know the codes, the better. I will also need that Uranium. From what I have seen, you do not have the correct fuel, and the Uranium 235 that you do have will have to be enough, which I why I will need so much. I will go through it faster than the correct fuel."

"I doubt that we can make the required grade of the fuel, at such short notice, so if it's the 235 that you need, then I'll get to making those phone calls."

"Thank you." He walked back to the top of the ramp. "If I contact the scramjet fighters, then please let me talk to them. They are very powerful, those planes, and I must get them fully under my control before I start talking of all of us working with Israel." He moved to his seat and sat in it. "There are a lot of displays that I am about to reveal. I would appreciate it if I could concentrate on those displays and find what is left of my crew, so if you have any questions, then please wait until I have finished." He looked at them all. "Do we agree?"

They all nodded – this had turned into his show, and they were not about to mess it up.

"Good." He flipped some switches, and a number of holographic displays appeared. The others just had to sit there, sometimes glancing at each other, but most of the time looking at the displays.

Although it was in the Homeworld language, they could tell immediately that it was very close to Hebrew. The tricky part was correctly understanding it, and sometimes they just had to give up. Avraham spoke very fast, and the responses were just as fast. He was obviously trying to identify where everyone was and then ask them to join him, but there were complications due to the distances involved.

They left him to it, as they couldn't have helped anyway. After an hour, he stopped and relaxed back into his chair. "It's done."

Benyamin asked, "What's the result? How many do you have, and can they get here? If so, when?"

"Let us go and find your Prime Minister and that Mossad man, Kaplinski, otherwise I will have to tell the story a few times." He shut down most of the systems, as he had done before, and they walked out of the pod and down the ramp. There was a security detail of three waiting for them at the bottom. He could see that they were Mossad agents, not Kidon.

Benyamin asked their leader, "Where did the Prime Minister go, and where is Admiral Kaplinski?"

"Sir, the Prime Minister has left, with Mr Kaplinski in charge. Should anything be needed, then he is the Man to go to for it. I believe Mr Kaplinski has set up a command post in the common room."

"He's gone? The Prime Minister?" The other nodded. "OK, no matter. Let's go and speak to Mossad, then."

They all left the hangar and went down the familiar corridors, entering the common room. Avraham went to the water machine. "I could do with a drink." He got a cup and then helped himself to five more glasses of it. "Anyone else?" They each came over and helped themselves to one glass.

Benyamin spoke to the Admiral. "Sir, our guest has a few things to update us on."

"Good. I was wondering how long you would take." He faced Avraham. "Now, sir, if you please, what information do you have?"

"First of all, no bad news." Kaplinski actually thought that his heart rate slowed on that news. "We've had enough of that today. There are some complications, which I believe we can get past. I have identified that some thirty scramjets survived. They are currently hiding off the coast of South America, together with the other pods. I have found close to two hundred more pods, with between one and five people in them. Although it was built for four, it seems that some people needed to be creative and make extra room. The pods allowed this, and so I have quite a few survivors, which I am very pleased to hear about. Altogether, it makes two hundred pods and some six hundred and fifty people. All of these people are highly trained military personnel, rather like yourselves. I have briefed them of my situation, and they are waiting for my signal to begin a journey here. Their fuel situation is like mine. They have some fuel in their pods, but they are running low. My hope is that once we have enough, then we can start the cleansing action centred around Israel and then slowly move it out. This is where the first complication arises. I will need fuel – a great deal of fuel. These pods are designed very efficiently, but their fuel supply is not indefinite. What I propose will require a large amount of energy, and the active pods will need to be replenished as they go. This is the hardest part. Do we know when I can get some of the Uranium 235, and how much?"

"The Prime Minister is still making the calls. We have some that is immediately available, and we are trying to assess how much we can actually put together. If we can't do it, then we will have to draw in some other countries, and that may take time."

"That is a fair point. We don't have time to waste. You will have to see the figures that Israel can come up with, and if necessary, I will have to take all that you can offer. There is another complication, which is that the pods are on the other side of this planet. When my ship was coming down, a lot of people got off at the start, which placed them in the South America area. They don't have the fuel to get here, not safely. Can you identify other nuclear facilities that can be approached for some U235?"

"That might be tricky. People are very particular about who they give their U235 to. But – we will try." He turned to one of his team. "Make enquiries with Chile. They have a facility in Santiago. Perhaps it has survived. Try Bogotá as well. There are a few in

74

France, England, and Germany. There is Russia, and that has a lot of facilities. We may as well start at the top – check with Russia first, then Santiago, then Europe. China has a few, but they may be too far from us to help."

"How can you know all of these facilities, off the top of your head? How do you not have to research these facilities?"

He smiled. "I am the head of Mossad. That is **Israel's intelligence and special operations service**. I make it my business to know – in detail – all of the possible enemies of Israel, and, believe me, when I say that Nuclear Energy, if abused, is probably the most catastrophic enemy of Israel. It is my _job_ to know these things!"

Avvy looked at this man, who was obviously deeper than he let on. "My apologies. I meant no slight on you or your people – but then you knew that…" He smiled. "I appreciate that what I ask of Israel may place it in an exposed position in regard to the rest of the world, but I must finish that with the proposal of what does it matter if I do exhaust all of your atomic material?" Kaplinski had to raise an eyebrow at this. "What if you have no more U235 at your disposal?" He paused, and they looked at each other. "What would you – Israel – do, if it knew that it had all of my technology at its disposal?" He stopped and looked at the man, and then turned to the others. "I am probably some five thousand years ahead of you in technology. At least!" He stopped again and slowly turned to everyone. "I do not say this lightly. I say this in deadly earnest – what would you do if I told you that I could offer you a cure for probably eighty percent of the world's biological diseases, and I could offer you a safety net – your 'Iron Dome', but more - a lot more – and I could offer a way for you to get to the stars? I need to know, and I need to know now. I will make it my business to ensure your safety, whether you have U235 or not. Would anyone turn it down?"

The head of Mossad had quite simply never experienced this before. He was a battle-hardened person who had seen many of his teams work and die in front of him. This? This was totally new, and part of his brain told him that everyone else in this hangar was going through the same searching. He thought about what he knew was going on in the world at that time, and what was going on in Israel, and what was going on between him and this self-admitted alien, this Admiral from another star system. This was not an easy choice to make.

"If you will, Avraham, I need to speak to the Prime Minister. Excuse me for a moment." He walked away to find a quiet part of the room, and he decided that he needed to get out, to have some fresh air. He walked through a couple of corridors, with the security detail always at his side, and he walked into the open air, next to the hangar. He raised his phone and dialled. "Prime Minister. This is Kaplinski."

"I know who it is, my old friend. What is the latest?"

"I am exhausted by this, and yet I am excited by it as well. This is unique, this is something that we have never had to deal with – that we never even thought possible! We have to decide, and within the next few minutes, whether we leave things as they are or whether we trust this stranger, this 'alien'. He has laid some very heavy demands on us, but promised us some fantastic rewards if we do as he instructs. The weight on my shoulders – on the shoulders of you and me – is not light. You and I need to be on the same page for this to work. We absolutely cannot disagree on anything about this. Michael, my friend – we sink or we swim on this next decision."

"Ha! Is that all?" He was tense but still laughing. Given all that was going on in the world at that time, it was remarkable that some people could even take a light-hearted laugh and get on with things in their everyday jobs. "Kaplinski, I made my choice when I was there. I'm sure that you did, too. We have no other options available to us. We must go with whatever plans this Avraham can come up with. You and I know that we have nothing better to offer. My advice, my old friend, is that we go with what we can see to be a firm plan, however crazy or mad it may seem. If it comes down to saving the world or dying, then I know which I will prefer, and I am sure that you will too." He paused, wondering how best to phrase what he wanted to say to this Man that he had known for the last thirty-five years. "Ram, we must do what we must do, starting now."

There was silence for a few seconds, as the head of Mossad thought things through. This was not easy. It went against everything that he had learned and taught over the last few decades. "Michael, you are, of course, correct." He smiled. It made it easier when you had an understanding of the position that you were in. If he were to do nothing, then they would most definitely lose everything. That much was guaranteed. If he were to do at least something, whatever that 'something' was, then they had a chance... a slim chance, but half a sixpence is better than nothing at all...

"Michael, you and I have a lot to do – I will not keep you anymore, but please make yourself available, should I call. Shalom, my old friend."

He ended the call, understanding that the Prime Minister would have a very busy next few hours talking to everyone that he could, and probably the international community as well – at least that part of the international community that he could trust. *So, to the next few hours of one Ram Kaplinski...*

THE MOVEMENT UNDER THE WATER MAKES A PLAN

The creature had moved through a number of water avenues and was approaching the southern tip of India. From his multiple sensory probes, he knew that the planet and all of the life forms on it were in very serious trouble. He could see that there was catastrophic damage in a line from South America through to the borders with China. Massive loss of life was obvious, and the casualty rate was climbing by the minute. He realised that there were so many things that had been damaged that he didn't really know where to start making any repairs. He worked it through that if he left the damage to the tectonic plates and recovered the air, then it would still fail due to the imminent collapse of many of the joins on the plates – the damage was just everywhere!

He sent more probes out across the seas, thousands of them, each one charged with identifying problem points and then doing their best to fix them. Once he had managed to stabilise as many of them as he could find, then he knew he could move on to the air quality issues, but he also realised that he couldn't do it all today. The repairs to the plates would take a few days, and he wondered if the active life forms could last that long.

With so much damage being seen everywhere, no one noticed that there were multiple shadows moving through the seas. No one saw that they sometimes registered, and even if they had, they would not have understood the significance of the signals.

THE ADMIRAL

BECOMES THE MOST IMPORTANT PERSON ON THE

PLANET

Kaplinski turned back and walked through the doors, almost racing to the common room. As he walked in, he announced, "People, listen up. The Prime Minister and I are in agreement, and the future of humanity on this planet is now passed over to Admiral Avraham, who is in all likelihood the only person on this planet who understands enough about what is needed to make this work." He faced the Admiral. "Admiral Avraham, you are in charge. You need to tell us what to do, what you need. Our options are exhausted. We go with you or we lose it all."

Benyamin was listening to this, and all that he could manage was a quiet "She-e-e-t!" The rest of his team said much the same, and a few of them said a silent prayer. "Oh crap, it's really happening."

Avraham was an Admiral in a far-flung star system, and he was used to giving orders to thousands of people, and he was used to having those orders obeyed. What was it about this new company that he found himself in? Why did he feel at home giving the orders, but almost reluctant to hear the response?

It took him but a moment, and then he began to feel the pieces fall into place. He knew only a little about the people receiving his orders at home – he was in charge of thousands of people – how could he know every detail about every single one of them? Even though he had been in this world for a very short time, he already

79

felt that he was 'at home', he felt at ease in the world, and he felt at ease with the people – they were the nearest that he felt he had to 'family'. That was it, he decided. He liked these people! These Israeli people – and then he had to ask, 'What is different?'. Why does he feel a kinship, a familiarity towards these people, when he could dismiss so many thousands of his home troops with an ordinary executive order?

He thought it through and realised that he liked them – all of them – and that made it different. These were not just names on a paper that you were sending into a battle or a skirmish; these were people, as real as he was. This is what it came down to. He was reluctant to 'play' with the lives of people that he knew personally and liked, but he could so easily 'play' with the lives of thousands of people that he had never met, and in reality never would.

He knew that he had a lot of things to do, and he knew that he had found the right people to help him achieve those goals. This was not going to be easy, but at the same time, he knew that he had to get it done. He must have a firm plan, and he must also have a plan for firm results. Anything else is just padding, just waffle…just a waste of time.

He stopped where he was and addressed the room. He knew what he had to say, and he also knew how to say it. This was a crucial moment. He needed everyone to support him, or he risked everything. He chuckled. If he risked everything, then so did the population of this planet. "Can we get everyone in here? Leave a security team outside if you must, but I don't want to have to keep repeating myself over and over."

Ram said, "We are working on that. We are also working on a video link with the Knesset, our government. From there, we hope to be transmitting to the United Nations at the same time." He walked up close to Avraham. "I am hoping that this will then be picked up by other networks and broadcast all over the planet. There will be some areas that will not see the message, due to the damage, but we will try to get as wide a coverage as possible. For better or for worse, we are in this. I have no idea what will happen at the United Nations. They are representatives of most of the countries of this planet, and they can get confusing. They each have their own perspective on things, and a lot of them are driven by their faith-based practices, and some, particularly the USA and Europe, are technically run as faith-free governments, even though they have people in charge who follow one faith or another." He sighed. "If it's confusing for us, then I hate to think what you will make of it!" He smiled. "Good luck. I hope that we can maintain the links, and I do really hope that everyone will listen. Understand that nobody else will know what it is that caused all of the damage. You will have to explain it all very carefully. Have you given any thought as to the language that you will use?"

"The language? Why no, I thought I would just use Hebrew, why?"

"I don't know how many languages you speak, but I was thinking that a different one might work better. Our reputation – the reputation of Israel – is not that good, in the International community, and for you to get up and tell the world in Hebrew that you caused all of the damage, but that you can put it right may really be too much for a lot of people."

"Well, I have a good understanding of some others, but not as well as Hebrew. I could do it in English, I suppose? When I first landed, I heard a lot of transmissions that were in English. I picked it up reasonably quickly."

Rab was impressed. "You picked it up by listening to radio transmissions? English would be perfect – it doesn't point the finger in any direction, so nobody can start to blame anyone else, certainly not us."

"English it is then. Thank you for your hard work. Let me know when you are ready for the transmissions. I have an idea of what to say, but how do I handle questions? Will they be able to ask me questions, or is it a one-way transmission?"

"It will be one-way at the start. I'll let you know when we have the full communications loop set up." At that point, another person walked over to them.

UNDER-THE-WATER REPAIRS

The huge, strange shape that had started from the Marianas Trench had approached the Philippines and moved around them. He was approaching Malaysia and decided that he needed to go through this small avenue of water called the Malacca Strait. It was the most direct option available to him. He planned to then go across to the southern part of India. After that, he would re-assess the situation. He was in constant communication with all of the sensory probes, receiving information on the activity below the surface of the water. The probes moved fast. Within hours, they had covered the Pacific Ocean to the east and had made their way to the western shores of the Americas. To the west, past India and approaching the East coast of Africa. He sent many further south, intending to send them around the southern tip and into the Atlantic Ocean.

He noticed that some water avenues went north, so he sent probes there, as well. They drove up through the Persian Gulf and others into the Red Sea. The damage reports kept coming in, on a scale that even he had not seen in his long lifetime. When the probes found damage, the repairs were immediately started.

THE ADMIRAL

TRIES TO TALK TO THE UNITED NATIONS

"Mr Kaplinski," he turned to Avraham, "Admiral. We are established with the links. We have the cameras over here, if you would follow me, please?"

They walked a short distance to a bank of electronics. "This is the area covered by the cameras. We are connected to the UN, and all that you have to do is stand in that area there, and look at that camera there." He indicated to Avraham all that he needed to know. "All you have to do is let me know when you are ready, and I will make it live. I will give you a thumbs-up sign, which will mean that you are on air. We have set up some monitor screens so that we can see them as well as them seeing you. If you need me to halt the transmission, then you just have to look at me. Good luck, sir."

"Thank you. I suppose now is as good a time as any?" Kaplinski shrugged. *This is your call now, Admiral.*

Someone else came to Kaplinski. "Sir, I have just had words from the Prime Minister. He has been making a lot of calls about this situation. He has managed to get Great Britain, Canada and the USA to agree. He is still working on the others."

He moved to the camera position specified and faced it. Everyone else kept out of sight. He looked at the camera while he had the details of what he was going to say going around in his head. He nodded to the technician, who did something on the bank of electronic equipment and gave Avraham a thumbs-up sign.

"Ladies and gentlemen! A small number of you know who I am, but many of you are still to have that explained to you, in detail. My name is Admiral Avraham. I am not an Admiral in any Navy that is of this planet that you call Earth." That brought a few rumblings from the crowd. "All of you are only too aware of the recent damages caused across this planet. Many of you have thought that it was a form of asteroid, and many of you have absolutely no idea what it actually was that came down. I can tell you now that it was no asteroid, nor was it any form of missile from anywhere on Earth. I am now able to tell you that what came down was the wreckage of my spaceship. I am an Admiral in a Navy that is a long way from here."

He could see on the monitor screens that the UN had erupted into a frenzy – everyone was standing up, arguing with their neighbour, everyone wanting to speak first. He paused, wondering how anything ever got done on this planet! With this sort of behaviour, it would take years just to agree who gets to make the next pot of tea! He walked closer to the camera so that only his head and shoulders showed. I need to get back in charge, I suppose.

He lifted his voice and shouted, "People – you, the people at the United Nations! Shut up! Now!" There was a pause as he waited to see if the crowd would begin to listen. "Sit down and listen to me, now, or you can all be blamed for the destruction of what is left of this pathetic planet!" He looked across to the technician, who pressed a switch.

"You are offline."

"Thank you." He voiced to Kaplinski. "I'm sorry about the shouting, but I need to be in control of this, or they will waste hours that we simply do not have. "OK, back on please."

"Now, the United Nations. Listen, and listen well. I do not have a lot of time to waste on you trying to sort out your own ego problems. I have a planet to fix, and I am running behind schedule as it is. I was trying to tell you that my arrival here is what has caused this damage, but I was also about to tell you that I can fix it. I cannot undo the damage caused, but I can prevent it from getting worse and wiping you all out! In case anyone does not yet have the figures, the air on this planet will probably be good for another week, but after that time it will be too polluted to use, and you and most animal life will then drop dead."

"I do not want that to happen. I will apologise to anyone who feels that they need to have an apology, but I will not – I cannot – be spending the next few days travelling around this planet, saying 'I'm sorry' to everyone, when I need to be organising how to stop this getting any worse."

He looked at the monitors and saw that he had them listening to him. Good. "I need to ask a few countries for some help. I require a large amount of what you call Uranium 235, and I need everyone who has the atomic processing facilities to let me know how much you have and where I can get it."

A lot of the teams at the UN started to laugh – *this was clearly some foolish idea, some person's notion of a joke, surely?* They all started to laugh, and it quickly turned to anger. *You want us to surrender our only deterrent against our aggressive neighbours? You expect us to give up our fuel supply for the next twenty-five years?* The rage in the statements was clear, but what it came down to was *'you expect us to believe you?'*

Avraham looked at the monitors and saw only fools.

"Very well. I will ask those countries that will help me to make themselves known to the following countries' representatives. Great

Britain, Canada, the United States of America, and Israel. These countries have already agreed to help me. Those countries that persist in not helping me will be the last to be cleaned. Your air will get worse, and in just a few days, you will have air pollution that will start to wipe out your population. We will eventually get to clean your air – we need to clean all of the air of the planet – but it is going to be done in a set of stages that I will be controlling by using some devices that have survived the crash of my spaceship. We will have to do an aggressive clean first, which will require a lot of energy. That energy can only be generated by atomic power, and your Uranium 235 is the best that you have that I can use."

By now, half of the UN delegations had walked out. Most of the Middle East countries, large numbers from Africa and the Far East. Those that were still there were Japan, Australia, Canada and a few others. Avraham wondered how many of those who had stayed even had atomic power. He looked at the technician, and he cut the signal. "You are offline."

"Thank you."

Kaplinski came forward. "That didn't go too well. I don't have everyone's details, but you have made the UN a right mess. Those who walked out didn't like the idea of handing over any of their U235 material to anyone else, especially not Israel. We will sort out the names, but I hope that we have enough for what you want."

"So do I. It is going to be risky from the start, and I must know how much energy I will have available. It is as if you were in a car going on a long journey, but didn't know how fast to go or how much fuel you had. Do you go slow, maximize your mileage, and take hours to get to your destination? If you could, would you go very fast, knowing that you had enough fuel, and get to your destination in half the time? If I go slow, does that give me time to clean the air, or does it just dirty up again the moment I have gone past?"

"OK, that's a fair point. Since the UN is now a joke of a circus, can you finish with whoever is left there? Tell them what's next?"

"Yes, of course." He signalled for the line to be opened, and got a nod from the technician. "To the members of the United Nations – those of you that are still listening to me - I will say 'thank you'. I did have high hopes for this 'United Nations', but I have to say that I am very disappointed. You are clearly not a 'United' anything. This will not be an easy time for any of us in the next few hours, so we must be in agreement. I will be contacting my people in South America, so it would be very good if the atomic stations there were on our side. There are not many to choose from, and it will be very important to know who will be in a position to push from there - who can I trust to be on my side? I don't know how best to explain the action, in terms that you will understand, but I will begin by creating a set of charged walls that will repel the dirt into space. It is a continuous process, and it will require the full attention of my teams, so it will be very helpful if I can at least have some guarantee of fuel supplies – if I fail at any point, then you all fail." He walked away from the cameras towards Kaplinski, and the technician cut the signal.

"Very well. If I am right, then this is the situation. A large part of this pathetic organisation, which is called the 'United Nations', has walked out. Is that correct?" Kaplinski looked over to the technician and the rest of his team, who all gave a solemn 'yes' nod. *Oh, terrific, my day just went from shit to ultimate crappolla. Great!*

"Yes, if I read the signs correctly, that's exactly what happened. We just gave away to the rest of the world that we could cure them of this latest plague, and they refused." He looked at Avraham.

"Admiral, please don't judge all of us on the actions of the majority. Because a lot of fools are in charge doesn't mean that those fools are any closer to the truth."

Benyamin stepped forward. "Avvy, please don't rush to any judgments. We in Israel are not fools – at least we don't consider ourselves fools – so we give close attention to whatever anyone else says – whoever they may be. We know that – and we accept that – we don't have all of the answers. What we have a problem with is coping with the trash that comes out of some of our neighbours, who say that they DO have all of the answers! They *cannot* have all of the answers! We know this, so why do they not know this?" He paused, wondering what to say next.

Avvy beat him to it. "So what you are saying, in as best a voice that you can, is that you believe that what you have to offer is better than what they have to offer? Is that it, in a nutshell?"

Benyamin was definitely caught off guard, that much Avvy could tell. "Er – yes, I think that's right. Oh, hell, we have to believe that we are better; otherwise, all of this, all of the fighting and dying for this country, is for nothing! Yes, we are better!"

THE ADMIRAL

HAS A PLAN

Avvy smiled. "Good. I made a very good choice in coming here, then. Now, shall we get on? I think I need to get to my pod now." And off he walked. "Ben, are you and the ladies coming?"

"Hell yes!" They all walked through the corridors into the hangar. As Avvy approached it, the hatch slid open, allowing the ramp to descend.

"I have my seat – the rest of you can fight over who goes where. Mr Kaplinski, once we are in, I will maintain a link to your systems. Ask me anything you like, when you like."

"Thank you, Admiral. Benyamin, ladies, please look after our guest."

"Oh, I intend to!"

Avvy sat in his chair, and Ben took the one opposite. The two ladies sat in the others. "Now, this is what is going to happen – " he shouted outside, "Mr Kaplinski, can you open the hangar doors, please. The time for secrecy is over." He returned to his crew. "I know this ship – you do not. I know what it can do; you will find out. I have a lot to do for the next few hours, so let's get it started, shall we? Don't forget the toilet facilities are basic, and down there." He indicated the next level down.

He started to power up the pod and identified the others that he had left at the bottom of the Persian Gulf. He powered them up and set their course to come to where he was in Israel. "I am starting up some more of these pods – craft exactly like this – and they will join

90

us here. I need a bit of a rush on this, so they will go sub-orbital and then drop down to us rapidly." He tried the radio, "Mr Kaplinski, can you hear me?"

"I can, Admiral. Loud and clear."

"Good. My other pods will be joining us in…" he ran through some quick maths in his head "in about an hour. They will be coming straight down, and very fast, so can you please tell your military not to try to shoot them down?" As he spoke, he closed the pod door, and the ramp hid away.

"Whatever you say. As an aside, if we are going to be working together on this, can we dispense with the ranks? My name is Ram."

"Excellent! Avvy will work for me. Now, Ram, can you chase up the U235? I need it in any shielded containers that you have, but I would like to have quite a bit. The fuel supply chambers on these pods are very sophisticated, and all I would have to do is put the containers in the supply chamber, and the system will sort out all the unlocking and loading, without anyone getting exposure to any of the radiation. I now need to take this pod outside to make some preliminary tests and change some settings. I can't do that from in here." The doors to the hangar were open by now, so he moved the pod slowly to them and out into the fresh air.

Golda asked, "Avvy, how much room do you need? Much of Israel is built up, with lots of small towns, so there isn't much room to play around."

"Oh, I thought that I'd go a bit further. I intend to go down to the Red Sea and find out what some of the damage looks like. I should probably stop at Mecca, if there is any of it left. That will allow me time to recalibrate everything. Oh yes, on the space ships that I know, and on these pods, we have artificial gravity. This will let me go very fast or stop very quickly without splattering anyone against the walls. You might want to sit down for this, anyway." He pressed some more controls, and all four seats had a full 360-degree presentation show, with gasps from the other three in the pod. All of this was new to them.

And then they flew – straight up. "When I said that we would go down the Red Sea, then I should have said over the Red Sea. If I stay low, then I will create a tornado as I go. I need to go up and then across and then down."

Tsipi asked, "Avvy, how fast are we going – it looks very fast!"

"Currently eight hundred miles an hour and accelerating."

"But we've only just moved a few seconds ago!" said Ben. Then he relaxed. All of this is going to be a surprise to all of us.

"And now we change course to go south, down to the Red Sea. Current speed one thousand two hundred miles an hour, and accelerating."

"Avvy, how fast can this go?" Golda asked.

"With the proper fuel – not this U235 – then it has a top speed of eight thousand miles an hour, when not in atmosphere. Within air, I must slow down, but since these pods have to deal with various types of atmospheres back home, they can have a range of top speeds. It is not my intention to kill any more people today, so I have to go this way."

All that Ben could manage was "Fantastic."

"Ram, I hope that you can hear this. I have to try a few exercises with this pod, which is why I am doing what I am doing. I will need that fuel when I return, which will be in about an hour. Don't forget about the other pods. They will arrive and either park themselves or just hover. I don't care about the spy satellites anymore. Just don't shoot them down."

They flew south, going over the Red Sea. They could see, on the displays, the huge smoke clouds in front of them, climbing up as high as they were and on, into space. "The force of the explosions was very severe. To have this kind of damage will take a lot of time to fix, and I don't know how much time we have left. I will be getting all of the pods online to mine, and then we will be using a screen that I can get generated. It's a variation of this tractor beam that you talk about sometimes. That is where I will be using the fuel. I will have to work out where to position the pods, and then initiate the screens. Once that has begun, there will be nothing else that any of us can do. It will be the computers and the pods that will do all of the hard work."

"Ram – can you get the word out. When I am generating this field, I will need the air to be clear – no planes or helicopters. Any that the screens find will be badly damaged, perhaps even destroyed. That won't apply to the scramjets as we use a totally different technology - the field generators on the scramjets let them go straight through."

"I'll get on it. We have identified an immediate source of perhaps enough U235 for you to work with. It should be here by the time you return. Our radar is tracking those other pods. They are coming in very high and very fast. I hope you have them under your control." He sounded anxious.

Ben voiced, "We don't like surprises. A few strange craft coming towards us fast are usually considered a threat."

"I appreciate that, but they are no threat to you, and neither am I. I will need them all to be loaded with as much U235 as they can carry. There is still a lot of work to do here, as this damage is spread over thousands of miles."

"I believe that I have a rough plan identified for North Africa and the Middle East, into Pakistan, and where the ship came down. I will need at least fifty pods for that, scattered along the line of the chaos. I will effectively set them up on either side, generate the walls, and hopefully push the debris into space. I will have to calculate a similar plan for South America, which will take another fifty or so pods."

"That's about a hundred pods – what are you going to do with the others?"

He laughed - a kind of bitter laugh, since he didn't know whether any of this would work. "Have you never heard of contingency? The other hundred or so are my backup devices. I need to keep this rolling until it's done. The pods that are in use will be replaced by the backup ones, and then the main ones will be replaced. This will continue until I run out of fuel or run out of pods. If I have run out of pods, then I do not know what we will do."

"What about those scramjets that you were talking about? What do you want to do with them?"

"They are another contingency. While the cleaning is going on, I cannot trust all of the countries to behave themselves. Their performance at the UN suggests that a number of them would even shoot down a pod if they could. My scramjets will be stationed along the lines, and any country that gets aggressive will have a few scramjets to answer to. They are more than capable of shooting down some missiles and then retaliating. Please don't get in my way on this. We do it my way or we lose it all. There is no other option. That reminds me – I need to get them more evenly distributed along the action line. They are all in South America at the moment."

"No arguing from here. We've come this far, and we all want to be breathing air at the end of this week, not choking in some Hell-hole smog."

"Good. Now, to action." He voiced some thoughts. "Ram, how about the U235 in South America? They will need it before I can start this."

"I'm not getting anything positive from them. They are as confused as anyone, and the idea of just handing over this material doesn't go down well. Sorry. We are working on the USA, though. They seem to be on board for this."

He wondered what to do, and then the commander in him took over. Opening a channel to the other pods and scramjets, he said, "This is the Admiral. There is a Uranium 235 supply based in Santiago, but they are reluctant to hand it over. Although I did hope for a smooth operation here, I must now ask five scramjets and a dozen pods to go in and get it. Try to avoid too much damage, but we are at a critical phase. If you can't get that U235, then all of South America is lost. It is not our fault that they are just too dumb to see that. You have a 'go' from me for this action. You must succeed! I will instruct my people here to negotiate with the USA people, and see where you can meet for an exchange of what they can supply."

Golda was listening. "Avvy, don't worry about it too much. You are, of course, correct. If you – we – need a great deal of U235 to make this mission succeed, then we must have it. The alternative is just too ghastly to think about."

"I don't enjoy it, but it must be so. I cannot fail! We have the means of putting this right, and the fuel is the only issue here."

Avvy sounded concerned when he stated, "Ram, you will have to make some political noises."

95

"True. Very well, let me do what I do best." Ram kept the line open, and they heard the conversation where he detailed the situation, as he saw it. The others could tell that it was not a straightforward conversation, but in the end, Ram came back to Avvy with the news. "He is not happy – hell, I am not happy! – but he agrees with the plan of action."

"It seems that the USA is well on the way to dropping a lot of U235 into South America, and Russia has agreed and is sending a lot down to Israel." He breathed out, sighing. "We may yet make this work."

Ben was delighted. "Russia came on board? Wow! Good for Russia! So how much do you think we can get?"

"If it all arrives in time, and we load it and make a start immediately, then we will need everything we can get. Russia is sending close to a quarter of a ton, which is fantastic. The UK has said yes, but is slow to get organised. Ram, I know that you can hear this, but I will take all that Israel can supply. The Russian supply is, at the moment, purely contingency. I need that material from Santiago! In terms of amounts, I will be using it all, and the Russian supply as well, I'm sure. Do not underestimate the scale of the work that I am proposing. This is not some Sunday afternoon run-around with your vacuum cleaners. This kind of operation has been designed and prepared, and we have run simulations, but up until now, we have never been forced into the position of actually using it to save a planet! This is new for me as well as unheard of for you. You must all appreciate that. Ram, have words with the Prime Minister – he needs to know this as well."

There was silence for a few seconds and then "Avvy, I realise that you didn't keep that from us deliberately, but I wish you'd said all of that at the start. The Prime Minister knows about the situation

– he has had an open link since he left the Dimona airport, so he knows everything that you and I have discussed."

"Very well. Prime Minister, can the Russian supply arrive at Dimona? I want to use it immediately it arrives."

"I will get them to make sure their route is with that in mind. We have some security clearances to make." Avvy thought he sounded tired, and then wondered when he had last slept.

"As far as I can make out, the pods that left the ship came down before your Buenos Aires was hit. They escaped on both sides and travelled quite some distance away from the line of descent. They are all safe, as the Pacific Ocean seems to have avoided most of the grief that the rest of the world has suffered. I have a number of them hiding in southern Chile, at a place called Pacheco Island, and some of the surrounding islands. Those are the ones that flew out of the starboard side and went south. The others flew out of the port side and flew north. They are currently in Lago Salinas Grandes and Lago Salinas Ambargasta. There is a third lake in that area, and the others are on a small set of islands that are in that lake, around something called El Mistolar. They are all secure at the moment, but their fuel situation is critical."

"Where is the USA going to deliver? What facilities are near?"

"I am hoping that they can fly into Cordoba, Argentina. It is central to the three lakes. I will then instruct all the pod commanders to meet the load at the airport. Time is of the essence here, and we can't start wasting it by being hidden. The pods will just have to be seen. I think there will be enough confusion around, still, so perhaps no one will care. My worry will be that the air disturbance will be too great. It is around five hundred miles between Cordoba and Buenos Aires, and there is a lot of bad air already."

"Send a pod to see. Make it one with the best fuel supply. Get them to fly high and see what the quality of the air is between the two places. If the supply plane can't get to Cordoba, what is your backup plan?"

THE ADMIRAL

MAKES AN ADMISSION

Quietly, he said, "I haven't got one."

"OK, try this. Get them to fly to Cordoba, and then along the line to Buenos Aires. They will have to stop at some point before it, because of the damage to the air quality. When they do, get them to go directly west towards Santiago. Find out if Santiago is still standing, and check their airport at the same time. If the supply plane can't make Cordoba, then perhaps it can divert to Santiago."

Avvy nodded, grateful for the help. He was on the radio, speaking for a few minutes. "One pod is going. They should be back with the information in half an hour or so. I need to check where the supply plane is. Ram, can you hear this? Can you tell me where the USA supply plane is, and what its route and arrival time will be?"

"They are currently two hours out, so they can switch routes between Cordoba and Santiago, if they need to. That exploration pod will need to be quick, though. That Russian supply will be within Israel in about an hour. They haven't wasted any time from their end!"

"That is good – that is very good. Can the plane or planes land at Dimona? Is the airport big enough?"

"Yes, that won't be an issue. What is your current fuel supply like, Avvy, and when can we expect you back here? I think you need to be here soon, so that your pods can be prepared and we can sort things out when the Russian supply arrives."

"We are done here. I have made all the calculations that I need. I think we will get there just about the same time as the other pods, so be careful not to shoot at either of us, please!"

With that, he turned the pod around and set it on a return course. There seemed to be little point in trying to remain hidden anymore, so he came back on a lower altitude and a slower speed, hoping that the current activity would all be worth it. *Would they make it?* This was a daring plan, and one that had never actually been used, not in a real-life situation.

He had to try. With all of the doubts and uncertainties, he knew that he at least had to continue with the plan. He was still alive and breathing, and would do everything in his power up until that last moment to keep the population of this planet alive and well.

THE ADMIRAL

MEETS THE RUSSIANS

As they returned to Dimona, Avvy parked the pod in the open, in front of the hangar. If he was correct, then he couldn't afford the time to park it every time and then have to get it out of the hangar. The world would know soon enough anyway.

He opened the door, and the ramp came down. The four of them walked into the sunshine and were met by a security team and Ram, who wasted no time getting them up to date on the changes of the last few minutes.

"Welcome back. I hope you managed to get your instruments set up with whatever calculations you needed to make. The Russian contingent is shortly to land, so I am curious as to where your other pods are?"

"They should be here – " and right on cue they screamed into view, coming down to the airport and stopping in an impossibly short distance. " – any time soon." He smiled, loving the idea that the plans were starting to fall into place.

They couldn't see the runway, but Ram then said that he had received a signal that the Russian planes had landed.

Ben was curious. "Planes? How many planes are we talking about here? I thought there was only one?"

"There is one supply carrier – the other four are accompanying fighters. The latest MIGs, so I think you would have some professional interest, Major? Russia is not prepared to send that amount of U235 anywhere by itself!"

"Good point. Can we get to meet the pilots? No, don't bother. I don't speak Russian, so unless we have a common language, like English, then it would be pointless."

"Major, you speak Arabic, Ivrit, English, and somehow you latched on to German. Why do you think that highly professional Russian fighter pilots would be any less intelligent than you are? They all have first-rate English, at least that is what I have been told."

"Fantastic! Can we go and meet them, then? It would be rude not to…"

"Of course, but first I want to know how we proceed with the U235 unload and reload into the pods." He turned to the alien Admiral. "Avvy, how do we do this?"

"I need the Russian crews to unload, and then I think I have to get at least two of the extra pods fully loaded." He stopped, working through how to get the logistics sorted out. "They will then have to go down to the west coast of Africa. I have very few pods in this part of the world. I need to get some of those from South America over to the west of Africa. I need to redistribute the locations and the numbers. I then need to get them all fully fuelled up before we start this. The USA supply, and hopefully what they can get from Santiago, will have to do to fuel up all of the pods that are left in South America. Those that come over the Atlantic will have to be fuelled up first, at their current location, and then make a very fast sub-orbital flight over the Atlantic. I will then need to get their fuel supply topped up from the supplies over here. I need them to be ready in West Africa as soon as possible."

"It's going to be ridiculously easy to explain, and a nightmare to action. I need to have half of the pods fully operational in South America. The other half must be here, spread between the west coast of Africa through to where the ship came down. I hope that it is enough."

"Each of the two theatres of operations will do the same. I need a line of pods on either side of the disturbances. They will all need to be fuelled up with as much as we can get. Once they are in position then I will activate the screens – this sort of force wall – that over the first few hours will, I hope, stabilise the rising clouds. Once that has been achieved then I will move into the next phase which is to push the debris out into space. It will have to go very high, and then I hope be picked up by your Solar Wind. This should then carry it away from Earth, at which point I don't care where it goes!"

"I need to balance the positions of the pods, as there are spots that will need to have a greater concentration of the force wall. It is no good to just put a pod every few hundred miles if one pod is doing all of the work and the ones on either side of it have nothing to move. The single pod is likely to overheat and burn out. This will be very difficult."

"I have left it with the crews that came down so that they can be aboard the pods or remain on the ground. Such is the style of this operation that I will not need the pods to be manned at all, as it is all driven from mine, although I will always need a skeleton crew who can do the actual cross-load of the U235. There is a sophisticated level of computer programming here that allows all of the pods to behave as one unit. If any pod experiences an issue, some kind of problem, then the ones around will be able to compensate in some way – to redistribute the workload."

He stopped and looked in turn at everyone. "The people that I came with, the ones who escaped from my ship, are all like me. They are the best of the best in all walks of life. There are designers, communication people, officers, canteen staff...and more. I don't have all of the details yet, but I would like to feel that, once this is over, then, if they wish to, they can make a home somewhere on this planet. Israel is my obvious choice, but I cannot speak for the others. I would ask that you start to think how you could absorb up to five or six hundred people into the Israeli way of life, if they should choose to follow me and live here."

Kaplinski smiled at the simplicity of the request and returned with "Admiral, don't worry on that point. We in Israel have been used to absorbing people into our country for a long time. We thrive on it, and the challenges that you and your people will set us will be met, and we will deal with them. It will be our privilege to welcome whoever chooses to make their home here. I am sure that they can be of immense help to us, and then eventually the rest of the world."

"Thank you. That is reassuring. Now, to business. Ben, you and the others can go and say hello to the Russian pilots, but don't be long. I want to be making a move as soon as we have loaded."

They left the area, wanting to meet with their opposite numbers. Russia and Israel had always had a tough relationship, and it was exciting that Russia had at least thawed in the face of total annihilation.

"Ram, who is in charge of the Russian contingent? I would like to speak to them and make sure that they understand the urgency of all of this."

"I understand that it is a Valerie Chekov, and he is coming to us now, from the supply plane."

Although the name meant nothing to Ram, even though he thought that it should do, as if on cue, the doors opened and some of the Mossad security walked in, accompanied by five people whom

Ram had never seen before. Before anyone could do anything, the leader strode forward and greeted Ram as an old friend. "Ram, Ram, Ram. How I wish that we had been able to meet under better circumstances. I have followed your career for some time. How are you, my old adversary?"

Taken aback by the welcome, all that Ram could manage was to say, "Adversary? I am well, thank you...and you are?"

"Ha! I am sorry. That is the Russian security. It really does work very well. If you don't recognise me or know me, then we have been doing some things right for the last few decades." He was grinning, and for some reason, a picture of the British actor Brian Blessed came into Ram's mind. Larger than life, loud, and always smiling. He clicked his heels together and slightly bowed in Ram's direction. "My name is Valerie Chekov. I am the head of the KGB - the real head that is, not the public one. Oh, Ram, I am so delighted to finally meet you in person!" He finally gave in and leaned forward and grabbed Ram, and gave him a huge Russian bear-hug. Oddly, Ram found that he didn't object.

He pulled back, and Valerie immediately wanted to know what the plan was. "Ram, I have to tell you that I am here after I have gone against a lot of my countrymen. Many of the Central Government didn't want to know about this strange plan to clean the air; however, it was to be done. I decided that, since I had some U235 spare then I would take it and let my Israeli friends use it." He stopped and looked around. "Tell me, Ram, who is the person who has damned this planet and yet may save it as well? I would like to meet this person."

Recovering from the bear hug, Ram turned to Avvy and said, "Admiral Avraham, this is Valerie Chekov. He has just announced that he is the head of the Komitet gosudarstvennoy bezopasnosti,

105

which translates as the Committee for State Security. It is the Russian State Security Department. Elsewhere on this planet, we always refer to it as the KGB. It seems that he has acquired some U235." He turned back. "How much do you have, may I ask? It is important that we know."

Unable to let the moment go by he said "Ram, you said that very well. 'Komitet gosudarstvennoy bezopasnosti' does not come easy to many people. Obviously the KGB is not unknown to you." He turned to Avvy. Snapping to attention he said "Admiral Avraham, I am Valerie. You may call me Chekov if you wish, but I prefer Valerie. It is both a shock and a delight to meet you. The shock is the circumstances, and the delight is knowing that we on this planet are truly not alone in this big Universe. That alone makes today worthwhile."

Avvy had watched all of this, the action and the dialogue between his new friend Ram Kaplinski and this newcomer, this stranger called Valerie Chekov. He had a disarming smile, this one, and a comfortable, confident air about him that Avvy was already liking.

Exploring what he thought the Russian would find acceptable, Avvy replied with "Valerie, I have only just met you, and if what I see is to be believed, I already like you. The people here are under a lot of pressure to get a lot of things done in a very short time, and any help that you can offer will be most welcome. I have briefly explained what needs to be done immediately to my friend Ram here, and then there is still so much to be done after that. Perhaps, Ram, we could hold a mini conference?" He stopped and turned back to Valerie. "Wait, did I hear you correctly, that you have brought some U235 without the agreement of most of your government?"

106

"Alas, that is so. They are not as quick as I am, and I truly see that we have perhaps hours to save Humanity, and all they do is bicker and whine about how their next few weeks are likely to be upset by the breaks in the food chain." He turned to Ram. "Ram, tell me honestly. I am getting my own data feeds, but I am sure that you have yours. If your government is supporting your actions, then I believe that your resources may be better placed than mine. I think we have perhaps two weeks at most before this situation becomes impossible to get out of. What does your information tell you?"

Ram looked at Avvy, Avvy looked at Ram. "Tell him the truth. He has already sacrificed a great deal coming here. Tell him!" As an afterthought, Avvy suddenly exploded, "I need to get to my pod - now! If what he says is true, then they will be coming after him. I need to make some changes." He quickly turned to Valerie. "Quick, how much have you brought?"

"One half of one ton. I would have brought more, but I had to get away quickly with what I could, or risk losing the lot."

"Shit. I will be in my pod. Valerie, Ram, can you follow? We must talk while I am on the move. It is most urgent."

Without waiting for further discussion, he turned and went to the pod. Ram signalled that Valerie should follow, and the pair almost had to run to keep up with Avvy. The ramp came down, and Avvy ran up the ramp. He went straight to the chair and set up the consoles and the views. The conversation that he had was hard to follow and very brief. He finished it and sat back in his chair. "For your information, I have managed to get ten scramjets to leave their current location and target here. They should be arriving within two hours, and that is only possible by their going into orbit and doing a powered dive to this location. I have asked one of them to get to the north of Israel, and to establish a position there."

Ram was curious as to what had prompted this activity.

"Because, if I read the Russian ideology correctly, then they will be really annoyed with Valerie, here. They will be mad that someone has gone against the wishes of the Central Government, and really, really pissed that they have succeeded. They are almost certainly going to send an attack force against us. They will not care about the casualties and won't care whether Israel survives or not. The fact that Israel fails in this endeavour is irrelevant. I have got to get the scramjets in place as soon as I can. With that in mind, I have to make sure that we monitor all of the radar, looking for approaching flights. If they come in, they won't be nice about it."

Valerie was quiet and then voiced, "Ram, he is correct at least in his assumptions. The moment that they realise what has happened, then they will be after me. I am sorry, I thought that I was helping, but I seem to be making it more complicated."

"Hardly more complicated than it already is. What is more complicated than the total wipe-out of Humanity? No, these are just some more things to factor in. Oh, and your question? How long do we have? We think perhaps three days before it becomes impossible to reverse."

He was visibly shocked. "Three...?"

"Ram, I must continue with the original plan. I can do a little with the few pods that I have, in terms of defence, but I must imagine that Russia will only send in some forces when they finally realise what has happened. I hope that it will take them a few hours to agree on. Until then, I must continue with the load up of the existing pods and the moving of some of them to the West Africa coast, to wait for the South American arrivals."

Ram appreciated the situation and turned to one of his team members. "Make sure that the supply plane is unloaded and then that the pods are loaded directly. Do not park the cargo anywhere on the runway - it must be straight off and straight on." The team assistant nodded and spoke to someone on his headset.

Avvy said, "Make sure that Ben and the others are there, and see if those Russian pilots can help, too. The smaller the number of people that is involved with this then the faster it will move forward."

Valerie wondered, "Perhaps we should go and see? This is a very delicate operation here, and I would hate to think that anything could go wrong. Not when we have so much at stake."

Avvy said, "I agree that you and I should go, as we will be better able to direct things from the supply plane, but I think that Ram should remain here. He is our main link to the Israeli Prime Minister and the outside world. We should not have those links interfered with in any way."

"Avvy, I will do as you say. I will keep all the lines open that I can. Can I assume that you are linked to your pod? It may be that the pod technology will be called upon very quickly, if those scramjets can't get here soon enough."

"All true. Valerie, shall we go?" They walked out of the room, along some corridors, escorted by the Mossad security team all the way. They emerged into bright sunshine, and they could see the pods and the Russian planes all grouped together.

They could see the Israeli and Russian pilots off to one side, and Avvy called out, "Ben, enough. We need all of you people to help with the unloading and loading." Although he didn't speak Russian, he assumed that Valerie had then called out the same to his people.

They all move towards the supply plane, Avvy and Valerie getting there just seconds after they did.

"Avvy, how should we best proceed?" asked Valerie.

"I need to see - how many tubes do you have? What is the way that you have packed this?"

"It is in easily handled small tubes - the weight means that a person can carry three or four at a time. We currently have a few hundred on board. I don't know the exact number as I had to grab what I could and run, but it will be around two hundred containers, each very securely packaged. One thing that we do well is look after our U235."

"That should mean that we can move it easily. Time is still not on our side. Ben, can you start by taking a look at the cargo and telling me how many you think there are?" Aside to Valerie, "Relax. This is to make sure that my people understand the situation first-hand."

Ben went into the plane and quickly came back out. "I'd say that it's around two hundred, possibly a few more. They are well placed, and we should be able to move them quite quickly and easily by hand."

"Valerie, I have five pods - six if you include mine. I need to load up each of them with an even distribution, except for two of them. I need more in those two, as they have to make a refuelling mission. I have more pods coming over from South America, and they will need to burn a lot of fuel to get here quickly. I will then need them to be refuelled at the coast of West Africa before I can safely start the main exercise of cleaning the air."

"Have I got you enough fuel? You will be needing a lot, I fear."

"I will indeed need a lot of fuel. I am trying to get as much as I can from South America before I start the cleanup over there. On top of that, if the pods that are coming here are fuelled up, then it means that what we have here can go further." He turned to this new ally. "Valerie, I will not lie to you, but I simply cannot have enough of this U235. If I run out, then it is likely that all is lost. You know the stakes as well as anyone, I am sure."

Chekov went silent and then said, "Let me make a few calls. A lot of people owe me a lot, and I intend to call in those favours."

He walked away and started talking into his mobile phone. That left Avvy to get on with the cross-loading of the U235 from the Russian supply plane to the pods. "I need pods one and two to have more than the others. If we have around two hundred containers, then I want thirty containers on each of the pods, and pods one and two get an even distribution of what's left. Is that clear?"

"Perfectly, but I don't know quite why, but I have to assume that you do, sir," one of the Russian technicians said, in perfect English.

Taken a little aback, all that Avvy could say was "You speak English?"

"Sir, we all speak English. And French and Russian, and Spanish. I can make a pass in Hebrew, but I am not confident in it, so English is best."

"Excellent! English it is, then."

He called across to the Russian. "Valerie. A question. Are these jets armed?"

"They are, fully. I don't travel light when I am with any U235. Why?"

"If the scramjets can't get here in time, then I may have to ask these pilots to go up against some of their own countrymen. Are they up for that, do you think?"

"First off, these are my people, my team. I believe that they will do as I ask. But, that aside, let me ask them." He switched to Russian and voiced his concerns. The same pilot who had just been spoken to answered Avvy directly, in English.

"Sir, I understand that you are an Admiral. I will assure you that Mister Chekov and I are one in this. You need to be given every opportunity to succeed; otherwise, we are all doomed. With that in mind, then yes, we will, reluctantly, go up against our countrymen. To not do this is suicide - and I have no intention of committing suicide, today or ever. Yes, my other pilots are of a like mind."

"Thank you. Thank you for your honesty and your commitment. Thank you from me and from the rest of Humanity. Let's just hope it doesn't come to it." On an aside, he said, "Ram, can we get these MIG planes refuelled? They may be needed sooner than we would like or not at all, but we can't wait until the radar says that we have incoming."

"I'm on it."

"Valerie, can we have a couple of these permanently manned, ready to go?"

"Of course. We will get them refuelled and manned immediately."

"Now, to business. Six pods, about two hundred containers. Ram, where is the Israeli U235? We could do with it now! I *need* the Israeli U235!"

"I'll need to check the latest activity."

Ram came back over the link, "It's on its way. Give us half an hour. We have matched Valerie with about a half a ton! Even I didn't think we were going to be able to do that!" He sounded happier than he had all day.

"Fantastic! That is a relief. Let's stick to the original plan - load up the pods with what we have. When the rest arrives, we will just double it up. Those pods for West Africa cannot take enough with them!"

The available personnel started to transfer the pods. The pilots and the Israeli fighter pilots did what they could to store the containers in places that Avvy specified. When they had finished loading, the Israeli trucks turned up. As with the Russian planes, the supply truck was not travelling alone - it had three trucks in the front and another three behind, each with four heavily armed Mossad operatives.

"Ram, the U235 has arrived. Perhaps you should be here to arrange this handover - they don't know me, and they might be suspicious of Valerie."

A voice from behind them said, "I thought that myself. I will do all the introducing and then get back to my post." Clearly, Ram was trying to stay ahead in the race to get the U235 sorted out. He moved quickly to the newly arrived convoy and introduced himself. Avvy thought that a lot of the people met him as if they were old friends, so perhaps they were, and it was wise to let him get on with things. Ram walked back, coming to Avvy and Valerie with three other people walking beside him. "Gentlemen, this is Admiral Avraham, and this is Valerie Chekov. They are both working with us on this." To Avvy and Valerie, he said, "Gentlemen, these are Moshe, Natal and Stefan. These are Mossad operatives, so they know me very well, and I know them. We don't need to have much explaining at this stage, as I have stressed that time is of the essence here. The containers are yours. What do you need to do with them?"

Avvy spoke first. "People, I need all of the containers offloaded and put evenly into the pods that you can see. I don't care what the numbers are, just divide them evenly between all of them. The Israeli pilots will be able to tell you where it all goes. Any questions, then ask me. So, any questions?"

No one had any. "Excellent. Then let's get to work."

All three leaned forward and shook hands briefly with Ram, and then followed it with shaking hands with Valerie and Avvy. They all moved away, back to the convoy, and began to unload. The people who had been loading the pods before stayed where they were and guided the new material into safe storage places. This took another half an hour, and by that time, with the full heat of the Middle Eastern sun, most of them were worn out.

The Mossad leaders reported back to Avvy that the supply was now completely off the truck and stored. The Russian pilots reported to Valerie that all was finished, and Ram reported that the refuelling for the Russian planes was underway.

Avvy wanted a review meeting, so he suggested that half of the Mossad team stay with the planes and the pods. He hoped that they would not be needed to do anything in terms of battle, but after all of the hard work then security could not be relaxed for even a moment. "Ram, I hate to say it, but the Mossad operatives should be ready to shoot first and ask questions afterwards. We don't have the time to waste!" He asked that Valerie pick a couple of pilots who would be ready in their planes.

"I need everyone else back into the hangar, and can someone organise some water? There are too many to fit into the common room, and I need to have a lot of water!"

They walked back to the hangar, Avvy now starting to get nervous about the situation for the first time. When they had all got into the hangar, he addressed them. The Russian pilots, the Mossad operatives, and the Israeli pilots. Ram, his team, and Valerie as well.

He addressed the gathering in English, which he knew fitted best for everyone.

"Thank you all for coming here today. Some of you have travelled far, and at great personal risk. It is my wish that I can reduce that risk to a more manageable amount, even if I cannot remove it altogether." He paused. "Why are we here? What has happened that has caused the Israeli Mossad and the amazing Israeli Air Force to work together with the outstanding Russian fighter pilots that are here, and a KGB leader?" He let his gaze fall around the room. "Me, that is why we are here. The incredible damage that has been done to this planet has been caused by my ship when it came crashing through the atmosphere. I apologise now to everyone, but it was not my intention to even be in this star system! I should be many light-years away from here, back at what I would call home, but I am not. My ship was too big, and it was built in space, and was only ever designed for use in space, so hitting an atmosphere literally tore it apart."

"I will tell you now that I have a plan to correct the damage caused. I cannot undo it, but I have a plan that should be able to stop it from getting any worse. That plan involves all of you here and some more of the people who arrived with me, who are at the moment in South America. It also requires all of the pods that I know about, which is roughly two hundred of them, and all of this fuel that

115

you have so bravely entrusted me with. For all of that enormous trust, I thank you all. I am trying to get the pods in South America to be fuelled up over there, with some supplies from the USA, as I need to have half of them over here, mainly in North Africa, in a line through to where my ship went down, in Nepal."

"I do not entertain any of this lightly. I will not give you any form of guarantee, because I cannot be one hundred percent sure that this will work. I can only give you my assurance, as a professional soldier in a Navy that is very distant from here, to you, other professional soldiers, that I will do everything in my power to make sure that some part of Humanity on this planet will survive. After I have done my work, there are still many things to be addressed. The damage to the food chain is immense. You will all have to face up to the idea that half of the surviving population of this planet may well perish before the next two years are up. The world will have to be on emergency rations for the next couple of years until Nature stabilises itself. I am sure that no end of what you call warlords will show up, each asserting their right to rule their little piece of dirt. There are many other political questions still to be addressed, and many, many more questions that right now I cannot even think of."

Ram interrupted him. "Avvy, I'm sorry to interrupt. You need to check something out. I am getting some odd reports, which, to be honest, don't make any sense. You may need to see what your instruments can pick up. Strange things, supposedly impossible things, are happening in various parts of the oceans."

Curious about this change in direction from his plan, he asked, "What sort of things are you talking about?"

"It looks as though when you came down, your ship set off some sort of chain reaction - a lot of reactions - along some of our fault lines, in the Earth's crust. We had so many other things to think about at the time that we missed a lot of signals - they were just drowned out by whatever else was going on around us all. There are a great number of minor earthquakes reported, all over the place - places where you'd expect an earthquake and new places, places where there has never been a quake recorded."

"I have not looked into it, but I suppose there will be some form of damage below the water. Do you have any idea how bad it is?"

"Well, that's just it. Various stations have begun to go over their reports, since the ship came down, and it is very bad - it looks, though, that the major fault lines are moving, or in some places giving way in large sections. This is probably due to the secondary impacts and the shockwaves that your ship caused. But here's the strange part. Some of the really bad breaks in the Earth's crust look as if they are mending, and that's driving everybody crazy, because they say that it can't happen! In the first few hours, they went from quiet to major fault, and they are now starting to settle down, at least in some parts of the world. And there's something else. A lot of underwater activity has been reported, which looks to have started out in the Far East, off the coast of Russia and Japan. The reports are blurry, but it looks as if a lot of underwater disturbance happened and travelled out from that region. Nobody has a clue as to what is causing this, but there is still a lot of underwater activity, still spreading across all of the oceans, and the damage to the fault lines in the crust looks as if they are mending. There is a latest report that some sort of underwater disturbances have been seen at the bottom end of the Red Sea, and that they are moving up, towards Israel. The same is reported at the bottom end of the Gulf of Oman, which suffered massive damage."

Nobody knew how to answer that, and eventually Avvy said, "Does that change what we need to do? I don't think so, unless anyone has any suggestions?"

No one had anything to add. "Does it look as if they are connected, these underwater disturbances and the repairs - are they in some way part of the same thing?"

"I'll ask our technical people. But on top of that, no, I don't think it changes what we have to be doing."

"Good. That was my thought as well. We shall continue to do what we can. Do we have any updates on South America?"

"I'll check that, too." He went away to ask questions of his teams.

Avvy continued to the crowd, "We all have a lot to do in the next few hours and days. It is absolutely crucial that we continue with the work; otherwise, we might as well give up now - and I have no intention of giving up. Not while there is a breath in my body. Much has already been asked of each of you, and much will continue to be asked over the coming few hours and days. We cannot rest until the job has finished."

"I hope that we can get some fuel in South America - we must! The USA is supposed to be on the way with an amount of U235, and I hope to get some from a refinery in Santiago, but this is a vital time being lost. The moment that the South America pods are refuelled, then I will be sending half of them to us, at high speed."

Someone interrupted, "What does that mean? What is your 'high speed'?"

A little peeved that someone had interrupted him, he volunteered, "They are currently over eight thousand miles away. If

118

they did one thousand miles an hour, they would take close to eight hours to get here. That is ridiculous. I propose that they will be launched into sub-orbital flights, and they should arrive in West Africa within two hours. After that, they will be refuelled by the two pods from here, which have the excess fuel, and then I will position them along the descent line."

"Two hours? That means..." and you could see that they were doing the maths in their heads. "That means that they will be running at four thousand miles an hour? That's - incredible." He was lost for words.

"Considerably faster. Four thousand is a conservative estimate." He looked at his audience. Time to be brutally honest, he thought. "People, think it through. Face up to what is happening around you, what the reports are telling you! If I come in at a speed that you can handle, then what is the point? You are happy, and the planet is lost! It makes no sense. I will not start this by lying to the Israeli contingent or to the very honourable Russian contingent. I will most certainly not begin this by lying to my Homeworld contingent. Those people have lost too much already - I am not about to add to that loss."

He looked around, and Ram interrupted to say, "We have just heard. The USA planes have landed at Cordoba, and even as we speak, they are off-loading. Give them an hour or so to get the pods stocked up, and then that group can start to get over here. I have said that they should concentrate on the West Africa pods first - get them airborne as soon as possible. While they are in the air then the people there can load up the remaining pods."

"That is great news! Amazing. I wonder how the Santiago trip is going. I need to check." He walked to his pod and entered the main cabin. Sitting down, he accessed the display panels and asked the questions. He took a few moments and then shut down the systems. He went back to the main group. Sadly, he said, "Santiago refused to hand over their supply. It was intact, but my teams ended up having to force their way in and take it. A few of the personnel in Santiago perished - about twenty. My people have extracted everything that they could lay their hands on and are currently going back to join the others. They have managed to get around another half a ton, spread over the few pods that went on the attack mission. They are all overloaded at the moment, and they need to get back and redistribute the U235 across the other pods."

"How long? How long before they can start their trip to West Africa?" from Ben.

"That will depend on how quickly they can get things cross-loaded. Nobody out there knows the urgency better than my people. They will be as quick as they can."

"Good. I am not doubting their commitment. Every second counts. Ram, can we get real-time updates from Cordoba, and will someone tell us when they have loaded up and started their journey across the Atlantic?"

Ben asked, "Where are the scramjets? Do you know how long they will be? I'd like to get to see one of them."

"While we have been busy here, they have been flying to get here. I think they are only a few minutes away."

THE ADMIRAL

INTRODUCES THE SCRAMJETS

On cue, ten hypersonic darts raced overhead, coming from the west and going to the east. Two broke away and moved north to cover Israel's northern borders. Three did a very tight turn and came into the airport. The others just hovered over the airport. The three landed, but the remaining five just stayed where they were, which reminded Avvy of the Israeli tactics when he was approaching Dimona. He smiled. Once a professional soldier, always a professional soldier.

The three jets approached the airport and came in at an impossible speed, and then parked just in front of the hangar. The pilots left the craft and started to walk towards the hangar.

"Ladies and gentlemen, please welcome three of my scramjet fighters." He looked around, putting on a very serious tone. "Please believe me that I hope that nobody here, or anyone that you know, will ever have to go up against one of these beasts." He looked most solemn. "You will lose." A pause, then "Everything."

The three pilots quickly identified Avvy and approached him. He knew the leader as they had served on a number of missions together. Speaking in Homeworld, he said, "Admiral Avraham, it is my pleasure to see you again, safe and well." He advanced and gave Avvy a huge hug. He whispered, "Good to see you, my friend."

"Commander, it is my pleasure to see you and your flight. You clearly made good time, and I see you have positioned the others well to keep up the guard."

"We are down here. The others will land when I say, and they will stay alert for any offensive actions from outside this sphere."

"Commander, let me introduce you to these *good people*."

The Commander lifted an eyebrow. *Did he just say that? He did, he just said 'good people'?* He must hold them in high regard, indeed.

Avvy knew that the Commander knew exactly what he had just said, but then the moment was past - it had been said, and acted upon, and they moved on with whatever was next.

Switching to English, he said, "Ladies and gentlemen, I wish to introduce scramjet Commander Ishmail, and his deputy, Commander Yitzak and pilot Yoni. I don't think they speak anything but Homeworld, so I may have to do some interpreting -" he stopped, as Commander Ishmail burst out laughing, and Commander Yitzak was having a problem keeping a straight face.

Wondering just what was going on here, Avvy asked, "Ish, what is so funny?"

Struggling to keep his sanity, all he could manage was "What would you like, Admiral? English, French, Japanese, or Chinese? There are five more I could offer, and yes, one of them is Hebrew. You are a very good linguist, Admiral, but trust me, you are not the only one in your fleet. My people and I have not been idle for these last few hours!"

Feeling miffed, Avvy managed, "I am sorry, Ish, I have been so very busy. I didn't think it through."

"Admiral, may I speak to these people?" Avraham nodded. In fluent English, Ishmail addressed the crowd. "Call me Ishmail. My full title is Commander Ishmail. I am the person who is in charge of the scramjet team that you have just seen fly overhead. I am from the same fleet that Admiral Avraham is from. My people are as one with the Admiral, and we agree that there is much work to be done. We, from what we call Homeworld, are effectively stranded on this planet, and none of the six hundred or so of us intends to be the only population for the next few centuries. We have learned some new languages while we were waiting, so we can converse with a number of people, from different backgrounds." Breaking out of English, he said, fluently, "Yes, we can handle Russian, too."

Chekov laughed and came forward. In Russian, he said, "Well done, sir. My name is Valerie Chekov. I am Russian, and I am here to give some fuel to the Admiral here."

He smiled back at the Russian, and switching back to English, he said, "You are the KGB man. I thank you - we all thank you - for your U235. It is urgently needed.

Avvy wanted to take back control and start things moving forward, so he jumped in with "Ben, take your people and go with my friend here, the Commander. He can give you a look at the scramjet. You will not be able to fly one, so don't even think about asking. Each is personally coded for their pilot. After we have saved the Earth, perhaps the Commander could take you for a flight - you and the others?"

The Commander was quick on the uptake. The Admiral had just allowed total strangers - these Earth people - into a scramjet. He must truly admire them, he thought.

"This will not be an open invite to everyone. At the moment, it will be an invite to a very select few - and here I include Ram and the Prime Minister, and you, Ben, and all of your flight. You will be most welcome, but it will all have to wait until we have settled the most urgent business of cleansing the air!"

Ben was amazed at what had just been laid out in front of him and the rest of his flight crew. By Avvy's own admission, he was some five thousand years in advance of where Earth was, in technology terms, and here was the Admiral saying 'come, have a look'. That was not something to ignore!

"Admiral, you and I have only known each other for a short time, and yet here we are, side by side, wondering how we are going to save Humanity."

The scramjet Commander was listening in to all of this. *How would it play out? What had these people discussed before he and his crew had turned up?* He was more than a little shocked by what was told next.

"Avvy, we have drunk water together, we have showered together, and my crew and I like you. More than that, we trust you. You are a very likable person, and you seem to have grown, in a very short time, to like us. We appreciate that. We appreciate who you are, and we welcome you, and we welcome your scramjet people to this tiny country called Israel. You are all most welcome."

All that the Commander could think was - Holy Klono's claws - *what have I missed?* Something has obviously happened between the Admiral and this Israeli flight crew - *what did I miss?*

Realising that he needed to get some more information, the Commander pulled Avvy to one side. "Excuse us for a moment, ladies and gentlemen. I need to speak to the Admiral on a few points."

He wasn't about to be fobbed off, but he realised that he had missed a lot of action between the Admiral and all of the other people.

"Admiral - Avv! What the Klono's claws is going on here? You and I have known each other a long time. I came here not really knowing what to expect, and I am confused! You have used the right phrases to me that suggest that you are one with these people, and yet you have only known them for a few hours! How is that possible, when you and I know only too well that this sort of link, back on Homeworld, can take years!"

Avvy laughed. He put on a formal voice. "Oh, Commander Ishmail. I do have a lot to explain, and not very much time to do it." He turned to Ben and Golda. "Would you two join us, please?"

The scramjet Commander was getting more confused by the second. *And now he invites them into our Homeworld private conversation?*

Ben and Golda walked towards the pair, wondering what was going on here. When the four of them were together, Avvy opened with "Ben, Golda, I know that you and I had an interesting time when we first met, in the air. You have to appreciate that Ishmail and I, and all of his team, are socially different from what the average person on this planet would expect. I know that you and I hit the point very quickly, and I appreciate all of that, in a way that you do not yet understand, and the likes of Commander Ishmail is struggling to come to terms with."

He was smiling through all of this. Commander Ishmail found this worrying, and the others were trying to work out just what was going on here.

Avvy finished the conversation with "My friends, there is much that we have yet to do. We will all shower together again, perhaps make love many times, and sit down afterwards, at a fine meal, perhaps get a little drunk, and most importantly, we will enjoy each other's company!" He laughed. "And now, to work!"

125

Commander Ishmail was in shock. All of this had happened, between the Admiral and these people, in such a short time? Truly, this was going to be an amazing planet to call home.

Avvy went to his pod with Ben, Golda and Tsipi. "I need to check out those reports that Ram mentioned. Something is rattling my brain, and I can't quite place it." He sat in his usual seat and pulled up all of the displays. He was lost in thought, and the others didn't dare to disturb him. It took half an hour before he raised his head and asked all of them. "Do you have old stories - very old stories - of myths and ancient Gods and suchlike?"

"We have thousands of them. Pick one - there is no end of them to choose from."

"No, I am looking for something specific. This would be a fabulous creature, a very old creature, and a very large one. It has probably had stories written about it, but never shown any real evidence. No one will ever be able to prove that they have seen it, but the tales are there anyway. These stories will be thousands of years old."

THE ADMIRAL

IDENTIFIES THE KRAKEN

They all looked at each other, shrugging. They didn't understand where this was coming from or where it might be leading. Tsipi voiced concern. "All I can think of is Godzilla, and that's of a Japanese origin. Avvy, what of it? What does this have to do with what we are doing today?"

"Nothing, and perhaps everything. I have re-read the reports, and I have made my own calculations. It is astonishing." He went quiet, wondering how this was possible. "The information that I can see almost defies explanation. Ram was correct when he said that somehow the faults in the Earth's crust are being repaired, but what I see makes no sense, unless I fall back on an old myth that we have on Homeworld. I have checked on your Internet and I think that you have - somehow - the same set of stories. As crazy as it sounds, both you and I have old tales about a huge sea beast, which you call the Kraken. It makes no sense to me to translate what we call it, so let's both of us call it the Kraken for now."

"The Kraken? You and us - we both have the Kraken?" Ben went off into thought. *How is that possible? That makes no sense...unless...* "That's crazy...that's...impossible! OK, here's what I think. We have some work to do. Urgent work that we cannot delay. There looks to be something else going on below the water, but we cannot delay what we are doing above the water. If we get diverted, then we will risk losing it all. As to the Kraken, well, we will have to see what happens, but right now, the Kraken, or something, is tidying up the fault lines and the seabed, and we need to let it. We need to be doing our stuff, as it needs to be doing its stuff. Let's not lose sight of the end game, here. I must mention that we do not yet know everything about the Universe, and I will not be surprised if the Kraken is a truly fabulous beast that is on all planets. How it got there - well, we will have to wait and see if it wants to talk with us."

"You are correct. We must start our work. So, Ram, where are we with things? Is the South American group ready yet? Do you know?"

"They are. The word is that they are ready to fly. How do you want them to make their journey? For what it's worth, I know the geography of the Mediterranean area and North Africa very well. I was wondering, if the pods have to make a line that is in excess of four thousand miles long, then there is little point in them all arriving in the West Africa area, say at Guinea or Mali. If they are all going sub-orbital, then have each one come down at different staging points, and we will send these few pods to intercept them there. Commander Ishmail, would you be able to secure Israel's borders with two or three scramjets?"

Looking at his Admiral, who just shrugged, he answered, "I believe that I could defend the borders with one, so three will be adequate. Why do you ask?"

"The pods that leave here will be overloaded with fuel for the incoming South American pods. They may be a little slow or sluggish in their responses. We cannot afford to lose even one of them. If we have the Admiral's pod remaining here for overall control of the situation, then there are five that can go to staging points. I wonder if you could assign a scramjet to each pod, to be safe? From what the Admiral has said, the scramjet should be able to handle any interference, which I hope will not happen, but we do need to be completely safe on this."

He thought about it for a second and then had to admire the logic. It made sense to have all of the pods looked after - he would have a word with the Commanders in South America and make sure that some scramjets came over with the one hundred pods that were about to launch. He also needed to make sure that they came down in groups from their hypersonic flights. That meant about twenty pods in each group, arriving along the long line from West Africa through to Nepal. He said all of this to the Admiral, who started keying in various changes to his many screens.

"It is done. There are one hundred pods matched on speed and course for West Africa. When they are at their full height, they will auto-adjust into five groups and come down at specific locations along the way. We will have our five pods from here, then meet them, and we will then start the cross-loading of the fuel cells. I have asked my people in South America, and I have three personnel on each pod coming over. That will make the fuel exchanges very fast. Once that has been completed, I will move the pods out into the line, either side of the disaster channel. When they are positioned, then I will turn on the screens, first of all, with a very low power setting. I need to get them perfectly positioned, run through many screen checks and then start to step up the power. In rough terms, the pods will take four hours to reach West Africa and the rest will take more time to get to their stop points - perhaps seven hours in total."

"As soon as the pods in Africa are loaded, even while the others are flying and landing, I will move the first pods into place. I don't want to wait until they have all landed and fuelled up before I move them. That will waste time."

"Once I have twenty-five in each line, twenty-five to the north and another twenty-five to the south, then I will start the screens. This will take quite a while, so do not expect quick results. I need to do this by our book, or I risk blowing up half of the pods through power overloads, and we don't have them to spare. It may take up to twelve or even twenty-four hours to step the power up to phase one. After that, it should be a little quicker - perhaps another twelve hours - but I cannot hurry this! Phase three will be reached very soon after phase two has finished. At that point, I will have nothing else that I can do - the machines will be in place and the screens will be pumping for all that they are worth to move the dust and dirt up, away from Earth. That will take as long as it takes, and I have no idea how long that will be. Certainly two to three days, perhaps even longer."

"Do we have that time? No, forget I said that. As you have just pointed out, if we try to hurry, then we risk losing it all. We will all have to make everything safe for the next few days, then. Israel may well be getting the gas masks out again."

"I'm sorry, I don't understand. Why do you have gas masks in Israel?"

"It's a long story. Let's leave it for the moment that some of our neighbours sometimes send over missiles that are loaded with poison gas."

Avvy was horrified. "Really? Are they that primitive? They are that stupid?" He was tired and beginning to show weariness. Anger crept into his voice. "Let us finish with this work. I will be having strong words with what is left of your neighbours afterwards."

THE ADMIRAL

GETS THE SCRAMJETS INTO ACTION

Commander Ishmail came online with "Admiral, I have just received word that Scram Five has just seen some missiles launched from what is called Iran. A place called Mashhad, or just outside of it. There are five missiles, all on a westerly course, with Israel their likely target. Admiral, scram five believes them to be nuclear warheads."

Ram came on with "*Shit! Damn, damn, damn!* I agree. We have just got that data feed. There is no announcement, so we don't know what their plan is. I don't know that Israel can get all of them."

"Damn it, why can't I get five minutes without this crap going on? Scram five, this is the Admiral. Time to go 'hot jets'."

The voice came back, "I am already on my way to intercept, sir. What are your orders?"

"Take them all down, wherever you find them. I do not want one missile in the air aimed anywhere close to here. Do we have more details on their exact route?"

From scram five, "They just launched. Direct westerly course, speed approximately one thousand miles an hour. I have approximately one thousand miles to cover before I intercept somewhere over their city of Tehran. I will be there approximately when the missiles reach it. Would you like me to wait until the missiles are clear of Tehran?"

"Ram, how many people are there in Tehran? Roughly."

"The central area has about nine million people. If you take in the larger area, then it's up to fifteen million or more."

"Ram, I can't wait. We can't take the risk of even one missile being missed."

Ram shrugged. "It's your call, Admiral. What would you do in your world?"

Avvy looked at the others, and they shrugged, too.

"Scram five, this is the Admiral. I need you to intercept and destroy the missiles - all of the missiles - at the earliest opportunity. Do you understand that order?"

There was only the slightest of pauses from the pilot and then "Loud and clear, Admiral. I am to intercept the missiles and destroy all of them at the earliest opportunity. Changing to battle speed now."

They didn't so much as hear the 'boom' as the scramjet went hypersonic, but they felt it, even though the jet was many miles away.

Ben asked, "What's 'battle speed', Avvy?"

"In this case, it's around two thousand miles an hour. He jumped from cruising speed to battle speed - there is no in between. That means he will intercept the missiles well before they get anywhere near Israeli airspace - he should get to them before they make it out of Iran, assuming that they continue on a standard course. He should have no problems getting all five, but Commander Ishmail, perhaps we could dispatch another scramjet as back-up?"

"Already on the way, sir. The back-up will stay a few miles outside of Baghdad, in Iraq, and take down anything that might get through. I will have a third on the Israel border, just in case."

"Ram, you might want to make some sort of announcement, or get the Prime Minister to. Anyone who launches against Israel will have us to contend with. *We cannot continue with these delays!* Whatever is launched, wherever it comes from, will be destroyed, and then we will back-track to the launch site and destroy that as well. We can do that quite easily, with our technology, and we do not take to threats very well at all."

"Scram five, when you have dealt with the missiles, track the launch path and destroy the launch site as well."

"I understand, Admiral."

He paused and then said, for everyone's attention, "I am aware that Tehran may be hit. It has already received considerable damage due to at least one of the ships' engines falling to the south of the city, into a lake, and causing a tidal wave. If one of these missiles lands anywhere close to what is left, then it may well wipe Tehran off the map completely."

"It will depend on the intercept location, which is only a rough point at this time. I am aware that perhaps nine million people are there. I am also aware that Israel has about the same number, give or take a million. I am not about to let the launching of those missiles go unpunished, and I need to make sure that the world sees what we are capable of. If Israel's neighbours launched fifty missiles, then we could take them all down and destroy most of the launch sites. We would reduce much of the Middle East to dust! The scramjets are not to be argued with!"

"It is possible that the missiles will not all travel over Tehran, but perhaps north over the water or further south over mainly desert. I find it hard to understand any mentality that deliberately risks their population by flying such destructive missiles over heavily populated areas!"

Ram wanted to change the gloomy subject. "Admiral, what's the latest on the launches from South America? I am aware that we are entering into many theatres of operations, and it is important to keep on top of every one of them."

Checking his displays, he said, "Correct. They are all at suborbital height and coming across the Atlantic Ocean. The escorting scramjets are keeping me up to date in real time. Give them another hour or so, and the first batch should start to drop to the West Africa coast. It's time we sent ours on their way, to meet them." He started to rapidly play with the consoles in front of him, and then reported that Commander Ishmail had taken charge of the escort, some of the scramjets going along with the pods as they left Israel and moved towards their destinations.

"Do you know where they are going to join up yet, or will it depend on the final descents?"

"I am close enough - five miles, give or take a little. It will take only moments to get them to meet up when they are all down. Right now, it is largely automatic. All we can do is wait. We wait for the pods from South America to arrive in West Africa, we wait for our pods to meet up with them, and we wait for Scram Five to take down those missiles. Ram, how's the Prime Minister doing with the announcement about the missiles?"

He came back with "Not very well. Some countries have listened, but Iran is not backing down. They have actually said that all of this destruction is some Zionist plot, hatched between Israel and the United States of America."

Avvy lifted his eyebrows, hardly believing what he was hearing. "Really? They said that?"

"Well, what they actually said was 'We, the peaceful people of Iran, will not tolerate being ignored by the world while all of this destruction is rained down on us. You only have to look to see that the USA and Israel, and much of Europe, are untouched by this destruction. It therefore must be caused by them, and we, the peaceful people of Iran, are prepared to strike back against this declaration of war against all of the peaceful people in the Middle East. The Zionist filth that is Israel must take the consequences of being aligned with the USA. That is exactly what they said, in an announcement that went out from the United Nations building just five minutes ago. So the 'peaceful' regime in Iran just launched five nuclear missiles against Israel. Nice touch that. Really 'peaceful'."

"Klono's claws! They actually called you 'the Zionist filth'? Wow, I'm beginning to get really wound up by your crazy neighbours. Once this is over, whoever is left is going to have a very rude awakening to the new way of the world. I would personally love to take charge of some diplomatic meetings. Scram five, what is your status?"

"Five hundred miles covered. I have the missiles on my screens, and can see that they haven't changed direction or speed. Intercept will be possible in fifteen minutes if I use the lasers. That might get three of them, and then I will take the others out when I get closer with the short-range systems that I have."

"What does he mean by 'short range'? What do you call 'short range'?"

"Anything up to one thousand miles, usually. He will probably be somewhere over the Iraq-Iran border and take them out with the short-range systems. The lasers can take out offensive weaponry at a much greater distance." Switching to directly address the scram pilot, he said, "Scram five, this is the Admiral. My instructions still stand. You are to engage and destroy all targets. I leave the how and the when up to you. Good hunting."

"Thank you, Admiral."

"While we wait for the first contact, can I get some water, please? It's been a hard day, here."

Golda said, "Avvy, let me get you some. You should have said earlier. We are just observers here. If you need anything, then you must just say. Do you want to try a little food? I'm sure this pod must have something that is from your Homeworld. It's probably a bit risky to try Earth food at the moment. We can't afford to make you ill!"

Ram intercepted with "Avvy, just a moment ago, you said 'Klono's claws'. I need to check something, but what does that mean in your world?"

"Oh, it doesn't mean much. Klono is one of those many mythical beasts that we have stories written about over many years, centuries. We currently have over two thousand different characters like that. Why do you ask?"

"I'm still checking...ah, yes, here it is. I thought the name was familiar. Admiral, in our world, we have many stories written over many years, and in general, they concern society and technology. Some stories are based on primitive societies, and some are projected far into our future. We have a category that we call 'Science Fiction', and many years ago, I used to read it. A *lot* of it. We have an author who used the expression 'Klono's claws' in a set of books that he wrote. Admiral - Avvy - what are the chances of Earth and Homeworld having two very similar languages, and then having two similar beasts like the Kraken, and then, on top of all of that, having similar mythical characters in a set of books?"

Avraham went silent for a moment, as what Ram had just said sank in. He heard it all, but just couldn't work it through. The odds of those combinations were millions - billions! Trillions? - against, surely? That cannot be a natural thing, which led him on to wonder just what was happening here, and just how this Kraken was involved.

Golda returned with a large jug of water, which Avvy drank all in one go. "I'll just go and get some more. Maybe we should get a lot of bottles on board?"

"Yes, a good idea. Ram, can you..."

"I'm on it. We'll fill your water stocks. I think that Golda is right, though. At this stage, we should not risk you having Earth food. We simply don't know how you would react to it."

"Alright, point taken!" He was getting irritable, and he knew it. *Neighbours that wanted you dead, mythical beasts that weren't mythical anymore...* He returned to the displays. "Now, let me check what is happening in the world today?" He checked through everything. "We have more disturbances in the water, just about everywhere. There is no impact to anyone, and it does look as if the shockwaves are receding, but I couldn't tell you why."

"Unless... It's the Kraken...?" volunteered Tsipi. "Face it, we just don't know."

"OK, we don't know, but I don't want to make guesses. It looks as though whatever is happening is working for our benefit, so I am not going to worry about it just yet - I can't afford to, and neither can any of us! We do have the first batch of the pods arriving in West Africa, so I will team them up with the first of our supply pods." He played with the controls again. "Now, how is Scram Five doing...?"

"Scram five, this is the Admiral. Status update, please."

"Admiral, I am about five hundred miles from the missiles. I have just taken out two with my lasers, and they have fallen. I think I took out their drive capabilities, so they would have just dropped. They are - what the hell?" He went silent, then came in with "Please hold, Admiral." There was silence on the radio, but Avvy knew the pilot well enough to leave him for a moment. He was obviously in a situation, and having the Admiral yelling in your ear was not a good idea.

Ram, however, did not understand this. "Avvy, what's going on?"

"Scram five is in what we call a situation. He has been interrupted while he was talking to me." *Please let him be OK.* "We have a standard set of responses for this, and the main one is that I leave him alone. He is busy, and I must not break his concentration."

"But what does that mean - wait, now I'm getting a feed." He went quiet, while his earpiece told him some news. He faced the assembled people. "Shit. Shit just happened. It seems that there are two mushroom clouds over Tehran. It's gone. Whatever your pilot did, he dropped the missiles right on their heads." There was another short silence, and then "Oh, shit. Oh, big-time shit. Avvy, what is your pilot up to? He should be able to see this. Tehran is gone, probably with at least ten million people. Ask him, dammit! What is he seeing?"

Avvy spoke into his systems. "Scram five, this is the Admiral. We have reports that Tehran is gone. Can you confirm?" Short and to the point. Always the professional soldier.

There was a silence for at least five seconds, but Avvy knew that the pilot was there. He came online with "Admiral, this is scram five." Another pause. "I have taken out all five missiles. My delay was in seeing that the first two that I took down went directly into

138

Tehran - straight into the centre. I was at a safe distance when the clouds appeared. I repeat - I am safe." Avvy could tell that the pilot was tense. Clearly, there was a lot going on where he was! Yet another pause, and Avvy was not inclined to hurry the pilot. Clearly, he was experiencing a mixed bag of emotions.

THE ADMIRAL

HEARS THE FATE OF TEHRAN

"Admiral, Tehran is dust. They have had a direct hit from two nuclear missiles, and there isn't a building left standing. It's at least a twenty-mile radius, and from what my sensors are picking up, you can tell everyone that they were dirty bombs. *Very* dirty. I have recorded what I can, but I can't get too close, and you had better tell everyone that this is now an official hot zone, and likely to be for the next five centuries. I don't know what the wind direction is, so you'd better get an alert out to the rest of the area. It's going to get hot for some time around here. The other three went down and exploded, but I have no idea what they may have hit." Another pause. "Admiral, we are professional soldiers, you and I. We have seen many die, often in our arms and certainly in our eyes. Believe me when I say that I take no pleasure, no delight from saying this. Tehran and its population are no more. Whoever is in charge of the politics of this planet needs to know this and acknowledge that this was not of my doing. Iran made to create an offense, and I took it down. It is firmly on the Iranian side where any blame is to be laid, not on mine, and certainly not on this country called Israel." Finally, he hissed, "*Admiral, you need to tell them that!*"

"Scram five. This is the Admiral. Good job. I would advise you to take what recordings you can and then turn around. See if you can backtrack and take out the launch site first, though, if you can. Come back to us. Come back to your new home, pilot. We need to talk, you and I, person to person."

There was another pause, and then the pilot came on with "Admiral, message received. If you don't mind, I will not rush back. I will set out for a cruise and see you as soon as I can. Admiral, you need to ask your new friends for more information on the capabilities of this country. If they have launched five nuclear missiles, what else do they have remaining? If they have any more, then you had better get another couple of scramjets on the Israeli eastern border, and now."

He sounded gutted - absolutely hollow - and Avvy wasn't going to push it. "Scram five, I will do as you say in spite of everything, good job. I look forward to seeing you back here. You have the location. At your convenience. Continue with the updates, please."

Avvy turned to his Israeli friend. "Ram, what other missiles and offensive weaponry does Iran have? Do you have the coordinates of anything that we need to know about? I want another couple of scramjets armed and on the Israeli border, just as Scram Five said. Ram, what else can we expect from these idiots?"

"We don't have all of the locations, but we do have some. I'll get the specifics to you now." He talked to his team and on his radio. "The information will be with you in just a minute. We know of at least a half a dozen major sites, but we have little idea of the number of minor sites that they may have. If they launch from one of their western sites, then they will only be minutes away from Israel. Can you take down anything in that short a time?"

"That will depend on where it launches from. I don't want to waste time with this, but if they launch anything else, I will be forced to go on the offensive. The scramjets are very capable war craft, and if we know where the launch sites are, then I will be forced to shut them down. It will be very messy for anyone caught up in it. Can you tell the Prime Minister to get hold of whoever is left in charge

141

in Iran, and tell them that if they launch anything else, then their country will be a dust bowl for the next five centuries. Tell him to find a military leader, someone who should have a little common sense, I hope."

"I'm on it. Here are the site locations that we know of." He handed over a few sheets of paper.

"Can I get this to scram five? He needs to be on the lookout for all of this. I'll send another scram forward to back him up. This other one needs to be on the Iran border."

"I think you'd be better placed if you stayed just west of Baghdad. That way, you can see a wider line of attack. Keep scram five where it is, and cover the attack line through to Israel."

"OK, good point. Yes, let's do that. Now, Ram, can I get back to setting up the pods, or is all of the Middle East going to get in my way? I can't afford to have these distractions. I need to be focused on the main problem of the day!"

An idea crossed through the very capable brain of Admiral Ram Kaplinski, and he knew that he just had to ask the question. "Avvy. Tell me. How much freedom do you give your pilots? If something else happens, do they have to report to you or can they make their own decisions?"

"In something of this type, it is usual for a very senior officer to be informed, and then he makes a decision - go or no-go."

Ram was smiling as he said it. *Would Avvy allow this?* "And you are the most senior officer from your fleet? But what if you have another senior officer, with years of practical experience, someone who knows the land and the people very well, but he is not of your fleet?"

142

Avvy finally realised what was going on. "Ram, you said earlier that you were an Admiral. You are suggesting that I hand over my scramjets to you?"

"No, Avvy, I would never do that. Don't be silly. I am not asking you to hand anything over to me. I am an Admiral still - I retain my rank until I decide otherwise - but I was thinking that if everything is left to you, then you will not be able to do the most important thing today that this planet has ever seen. Look, you won't know about it, but we have a television series called MASH. It's to do with a medical team in what we call the Korean War. I remember things. I remember odd things. There is a line in one episode where the nurse says that they could transplant a vein and save the leg. The doctor says 'We will save the leg, but lose the patient!'. It's like that here. We need you to do what you are best at - what *only* you can do! Everything else is just interference. If we wait, then with you in charge, we will win all of the battles - but we will *lose the war!* That cannot happen."

"If you agree with your pilots that they update me, then I will make the decisions, without interrupting you. If it is an impossible situation, then the scramjets may have to call you, but for the next few hours, you need to be single-minded on setting up the pods. I can take control of the Iran actions - and anyone else's - and advise the scramjet pilots. Scramjet five has had a rude awakening to some things that can happen on this planet, but he is good; he is a professional. If Iran, or if anyone else, launches, then you do not need to know every detail. We can handle that, if you will allow us."

He thought about all that had been said and wondered how he had missed what should have been obvious. *Perhaps I am too close to the action, perhaps I need to break away from all of this.* "Ram, you are correct. I must step away from some things and get the pods operational." He spoke into his radio. "All scramjet pilots, this is the Admiral. I am bringing a person from this planet on board to head up a part of our operational team." Half of the scramjet pilots raised an eyebrow - *really?* "As from this point on, scramjets are to liaise with Admiral Ram Kaplinski directly. I will set up the radio connections. He is to take immediate charge of all operations over Iran and Iraq, and he is the one to go to for advice. If anyone in the Middle East launches, then update Admiral Kaplinski, as you would normally update me. I have other, *more important* work to do now, and this Iran nonsense has delayed me. All scramjets acknowledge, please."

Each scramjet pilot went through a short acknowledgement back to the Admiral. One even managed a "Welcome aboard, Admiral Kaplinski, sir."

"Good. That is done. They accept you, Ram, on my say-so. They are the best, and I have to acknowledge that I believe that you are among the best that this planet has to offer. Look after them, and they will look after you. Now, can I please get back to what I am *supposed* to be doing?"

Ram came in with "People, now listen up. We have wasted enough time being distracted from the one thing that we must get started on today - now! As of now, if anyone has questions or needs anything, then they should come to me first - Admiral Avraham must be allowed to do what he started on, with no distractions. Is that understood by everyone?"

Everyone nodded in agreement, and Avvy said, "Good. So let's start."

He asked Golda, "Do we have that water onboard yet?" She said that it was, and he said, "Good. So are you three ready for the trip of your lives?" They were all ready, and without any further delay,

Avvy said, "Ram, my scramjet pilots are yours. I need to be in the air for this. I will not see you for some time, but we will keep the link open. I need to get this thing started."

"Have a good trip, Admiral. I will see you on your return."

Avvy closed up the door to the pod, and it lifted off, straight up. Since he was still positioning the other pods in South America, he now had to add in the ones that were along the few thousand miles of North Africa into Nepal. To the others in the pod, he said, "This is a tricky part, where I have to jostle all of the pods. I am still trying to balance the fifty in South America, and I will then have to add in the ones over here. I need to position them first and then add them in, one by one. I can't just add in another fifty pods. That would start to blow circuits, and we can't afford that. I'd appreciate it if you could not interrupt, but just nod every now and then when I give you any updates. This will take a few hours, so settle in. I should have told you to bring a good book or something."

"Never mind a good book! We are on the trip of a lifetime. That will do, very nicely!"

Avvy went back to the displays, blanking out the others. They could see that he was incredibly busy, but they had no idea what he was up to. After a while, he would ask if there was any water, and Golda would get one of the many bottles. Without taking his eyes from the displays, he would take the bottle, say a short 'Thanks' and down it all in one go. This whole business of setting up the pods went uninterrupted, and for that, Avvy was grateful. At the back of his mind, he appreciated that his passengers were letting him get on with the one important thing of the day. It took a few hours, but he then relaxed and sat back in his chair.

"It is done. They are all in place, powered up, all connected, ready to go into the next phase. Right now, I need ten minutes or so for a break." He got up and started to walk around the limited space in the pod. "I need to use the lavatory facilities. I'll see you in a few minutes." He went down the stairs, and they all stayed where they were.

Benyamin said, "Hey, did we bring any food for us? All of a sudden, I'm starving. I've been so focused on what Avvy has been doing, I forgot to eat."

Golda said that it's been brought on board and stored in various locations. She pointed out where the nearest batch was. "Go help yourself. Don't make too much of a mess."

He went to find the food. Golda and Tsipi looked at each other. "You like him, don't you?" said Tsipi.

"Avvy? Of course, I like him. He's very likable!" She smiled, wondering what would happen when they got through this and the two of them finally ended up in bed together. "I see that you and Ben have hit it off, too."

"Yes, it's true. After all of this, I may have to decide on what to do. Should I stay in the Air Force, or is it time to settle down? Benyamin will make a wonderful father and husband, I am sure. We will discuss this after we have finished here - assuming that this all works and we can have a future at all!"

"Yes, that's a good point. There is so much that depends on this crazy plan working. I want it to work, I want to have a future..." she trailed off, lost in deep thoughts about how fragile human existence had become in such a short time. Israel had always had a fragile existence, but now the scale had completely changed. Not just Israel was under threat - now it was the whole planet, and the future of Humanity that rested in this crazy plan from this alien who had arrived and caused the problem in the first place! She had to laugh, which surprised the other.

"What's so funny?"

"Here we are, all worried about the next few hours, hoping that this works. It's not as if any of us get out of this alive, at the end, anyway!"

They both chuckled at that as Ben came back. "I brought you guys back some stuff, just in case." He handed over some pre-packed food and water, and they started on it. "Thanks, I guess we all got too engrossed in what was going on."

Avvy returned and smiled at his new friends. "Are you people OK? I'm sorry if it's a little boring, but I don't have any in-flight movies on this pod."

Golda said, "Avvy, we are concerned that you are OK. We will make do with what we can find. Do you have time to have some of your food? I don't know when you last ate anything."

"I'll take a little something now, and another water." He went to where the Homeworld food was stashed away and got a small package out. Returning to his seat, he asked, "Are you OK to stay with me? The next part is going to take quite a while as well, and I can't afford to hurry. If you wanted, I could drop you back home?"

Golda answered with "Ha! Are you kidding? And miss out on this? Seriously, if we are at all getting in your way, then we will go back home, but if we are not getting in your way, then this is the place that we want to be. We want to be able to say to our grandchildren that we were there when this mad alien Admiral came and saved the world."

He knew that she was trying to lighten the mood, and smiled back at her. "Thank you. Golda, you and I should have a long talk after all of this. I have much to still learn about this planet, and I would appreciate a good teacher."

She smiled back at him, "Avvy, I will check my busy calendar. I am sure I can make some time, somewhere over the next few years. You and I will make the time, of that I am sure."

"Good. Thank you." He addressed them all. "Thank you, all of you. You have taken a lot on board in the last few hours, but you have risen magnificently to the task. I could not ask to be in better company."

Golda came over to him, rested her hand on his arm, and smiled. "Avvy, that goes for us, too. We are all delighted that we have met you, in spite of the circumstances, and we all look forward to a long life together, knowing that you will be part of it."

He gently placed his hand over hers. "Thank you, Golda." He looked at the others. "Thank you, Benyamin, especially for not shooting my pod down, and thank you, Tsipi. Thank you all for wanting to be a part of this."

"And now, back to business. I will run through some more checks with all of the pods and then start the next phase, which is powering up the screens. This will be tricky. I have one hundred pods, all in position, at cruising power. They are just in position. I have to bring the screens up and get the power slowly stepped up to a workable level. It will be slow and dreadfully boring, I am afraid. If you need to rest, then I suggest that you go downstairs, as there are a couple of pull-down cots that two of you could use."

"I don't know about the others, but I find that these seats are very comfortable. I could easily sleep in this," said Ben.

"OK, sort out whatever suits you. I need to get on..." With that, he turned back to the displays and went quiet. His concentration was obvious, and nobody was inclined to interrupt him.

THE ADMIRAL

POSITIONS THE PODS, AND IRAN TRIES AGAIN

While all of the positioning of the pods was going on, two things happened elsewhere. One event was a set of missile launches from various locations in Iran. The ruling authorities had clearly not learned from the earlier events. A barrage from Semnan flew upwards and turned sharply west, towards Israel.

Scram five immediately came online with "Admiral Kaplinski. I am on the Iran-Iraq border. Multiple launches from Semnan. Number unknown, but at least thirty-five. I will take out what I can, but I need scram backup now." With that, he was gone. He raced towards the incoming missiles, taking out what he could with the long-distance lasers. Half of the missiles fell into the desert, and at least three landed in the middle of Qom. They all then exploded, obliterating the area. He was picking them off one at a time when another scramjet came on his radio.

"Scram five, this is scram three, don't sweat it too much, but I have your back. I am currently four hundred miles behind you, advancing on your position."

"Thank you. I will maybe leave you some." There was a light touch to his voice, even though he was concentrating on the action at hand.

"Very generous of you. I need the exercise. From what I can see, the pack has split. You take the southern group, I'll take the ones to the north."

150

Scram five wondered why they had split. "It seems odd to split the pack. Perhaps they did learn something from the last time, and they hope that at least something will get through. Time to take them all down."

The action was short and very brutal to Iran. Every missile that landed crashed in a mess and then exploded. In a way, it was lucky that these were not nuclear warheads but more conventional ones. Thirty-five nuclear explosions would have trashed everything for thousands of miles around, making most of Iran and the Middle East a wasteland, but the more conventional warheads were still capable of inflicting massive damage. Many of the missiles landed in the desert, but the cities of Isfahan, Najafabad, and Masjed Soleyman all suffered huge casualties. One missile, the last to be downed, even went as far as Shiraz, in the south.

Ram had to speak to the scramjet Commander. "Ishmail, your scramjets are amazing. Do we have air cover over Iraq? I need to send those two scramjets on a search and destroy mission, but I can't afford to weaken the Israeli borders."

"Ram, no problem. I will be back in the air in two minutes. Send them the mission details, and we will be the new border patrols." He was gone, racing for his scramjet, along with the other two that had landed with him. They were in their cockpits and then whoosh! They were gone. "Scram five and scram three. Scram one is in the air. I have to ask you to backtrack to the launch site and take it out. Can you do that?"

"This is scram three. We are on our way. They will regret this latest set of launches if anyone actually survives our arrival!"

With that, they were gone, at a ridiculously high speed. Within minutes, they were in range and both had targeted the launch sites. They fired their offensive weaponry and then turned for home. They knew that Semnan would not be firing any more missiles for a few years.

"Admiral Kaplinski, this is scram five. Targets are destroyed, and we are returning to the western border. If you have any more likely targets, then we should inform any Iranian leader that they should stay quiet, or we will go on the offensive - I see no reason for us to sit around and wait for the next launch. We can easily take the fight to them! Admiral, find someone, talk to them, and tell them that they should behave, or we will chase them down - all of them! Their casualty list might well destroy what is left of this damnable country!"

The two scramjets went into cruise mode and flew comfortably back. They were only slightly inconvenienced when a couple of surface-to-air missiles came up at them, but they didn't even need to alter their course. Their lasers took them out immediately, and they launched back at the site that the missiles had come from. They saw a small puff of cloud as it was destroyed.

They reached the Iran-Iraq border without further incident and set to hover. Scram five updated Kaplinski with the situation, and then they both just stayed there, waiting to see if anyone else in Iran fancied their chances at taking down a scramjet. After two quiet hours, they decided that various people had rethought their plans.

The other event that happened during this time was less of an 'event', but more of an observation. The world's instruments were back online and measuring what they were supposed to measure. Part of this was now helped by the fact that the human observers had stopped running around like headless chickens while they tried to figure out what was going on. Not a one had actually reached any firm conclusion on anything in the recent hours, which meant that all of the amassed data from the time that the ship crashed through the atmosphere had generally been ignored. Now people have started to go back to it and look at it. And they were horrified by what they saw.

Ram was in close contact with the Prime Minister. "We can't understand what we are looking at. It is there, in front of us, but it makes no sense. We can see thousands of isolated underwater activities, which on the face of it cannot be happening, but clearly are. That hurts. We don't understand it. The underwater faults look as if they are mending, which is insane, by any standard that we know of. Having said that, if someone had said just a few hours ago that we would now be assaulted by an alien starship and then be busy trying to stave off our own annihilation, then I doubt you would have had any takers. We know that something is going on, we just don't know what it is!"

In a very weary voice, the Prime Minister answered him. "Ram, don't sweat it. Whatever it is looks to at least be doing something positive. We will have to take each hour as it comes. There is too much to worry about today, without making any more stress."

"What concerns me is the recent update. There are movements under the water, and they are moving, very slowly, up the Red Sea. What happens when it gets to Sharm El-Sheikh? Does it split and go up both of the channels, or does it stay there? Whatever it is!"

"What do we know of this thing?"

"Nothing apart from the size. It's enormous. Probably a half a mile wide and perhaps two to three miles long. We have no idea what it actually is!"

"Update Avvy when you can. We must make sure that we have covered all the bases on this. It is certainly a day that no one will ever forget!"

THE ADMIRAL

FIRES UP THE PODS

Avvy had blotted out the others and concentrated on the work he needed to do. He had made constant adjustments to the one hundred pods, balancing them and re-balancing them all the time.

"OK, people, here we go with an update. I know that you are listening, Ram, and I assume that the Prime Minister is as well. We are going for the main part of the exercise. I have put them all online. They are all connected, and they are at a prearranged altitude. Given that Iran and possibly others may still decide to be a nuisance, I have moved all of the pods to a high cruise altitude. Twenty-five miles should be outside the range of most missiles, and I have to trust that the scramjets can get to any ICBMs before they do any damage. The screens are activated on minimal power. I am now going to slowly increase that power. This, again, will take some time, so the sooner I start, the better. We are positioned above Sharm El-Sheikh, so that I can see the operation better, some two miles up. Try to keep all air traffic from our area."

Ram came in with "Avvy, you are free to commence with your strategy." He hoped that Avvy would not ask too many questions, but Ram left it with, "We are in control of Iranian airspace. The scramjets are incredible."

"I know. Right, beginning the next phase. A slow power up to the screens." With that, he was gone, concentrating on bringing up the power in each pod and making sure that the overall effect didn't rest in just one or two, with the others not doing any work. It took him the next five hours before he relaxed back into his chair and

154

said, "It is done. They are all active, with a full power setting. This may take a while to actually present itself, but when it starts, you should be impressed." He smiled. "Golda, any more of that water? And a little food? I don't think I have anything left to do now except wait, and hope it all goes according to plan."

"I'll get the water and food," and she was gone and back in seconds.

"Thank you. You might all need to sit down for this next bit." They all duly sat down, wondering what was about to happen. "Shall we see what is happening outside?" He played with some more controls, and the image of the inside of the pod vanished, to be replaced by a clear view in all directions.

They all gasped and looked around. Everywhere was clear - where had the pod gone to? "Avvy, this is amazing, but what is it? I know that we are still in the pod, but I can't see it."

"Relax. It's another part of the technology. We will get to discuss it later, but for now, try to enjoy the show."

They could see everywhere, in all directions. Avvy explained that on the arm of their chair was a ball control. "Move it around, and your chair will move in that direction. You are pilots. You'll figure it out."

"I needed to get some height for the visibility. This allows me to see the effects as they happen."

They looked over to the south and could see that there was an effect, almost like a heat ripple across the whole sky. It ran from the left as far as the eye could see and went across to the right, disappearing off into the distance. It started at the ground and went as high as their eyes could follow.

"How high, Avvy? How high does it go?"

"My instruments say that it is in excess of one hundred miles. It is more powerful down here, at the sea level. This is where it has to do the most work. Towards the top, it starts to fade. I hope that is enough. It should start soon, and then you will see the fireworks."

Tsipi asked, "Fireworks...?" and then it started. From the ground level, there was a pulse that moved up through the force screens, rushing up towards the sky.

"Yes, fireworks. Don't panic, but there will be lots of fireworks, and it will go on for some time. I don't know for sure, but given the current dust cloud pollution, there is a lot to clear up. This could go on for hours." He looked at them. "Or days. I did mention that you should have brought a good book."

Ben asked, "But are you saying that we have nothing more that we can do? It is now automatic, and we just have to sit and watch? For hours?"

"Yes, that's it. This part of the operation is not something that can be done in ten minutes. The force screens are aligned away from each other, in a slight 'V' formation, and hopefully they will contain large parts of the debris that has been kicked up. Once the screens have established themselves, then they will continue to push the junk upwards, where I hope it will be sent into space, and carried away by the solar wind. Once it reaches its optimum speed, then I can do nothing except let it run. There are millions of tons of dust and dirt that have to be removed from the air! This exercise will take as long as it takes. I'll take off the all-round view for now." He did something to the controls, and then the pod control centre was back.

Ben said, "OK, ladies, since we have nothing that we can do for some time, I am proposing that Tsipi and I go downstairs for a while

156

and get some rest. Golda, would you mind looking after our friend for a few hours? You can come and wake us up if you like, but I think we need a bit of a rest. Tsipi?"

She had realised what Ben was saying and answered with "Yes, I'm tired enough. A decent sleep will do me the world of good." They both got up and went down the stairs to the lower level. Golda let them go for a few minutes and then said to Avvy, "If you hear any funny noises from down there for the next half an hour, then don't worry about it. They will be entertaining each other, and probably sleep afterwards."

"Entertaining?" And then he realised what they were talking about. "Oh, you mean sex! Why didn't you say? Oh, Golda, I am not that bashful. They didn't have to go away if they didn't want to. I am quite sure that we could have enjoyed their company while they were busy."

Golda wondered about that. "I know that we have discussed this, but it's possible that they did want some privacy, even for a short while. Avvy, in the past, we have all been busy with each other, in the same room, while we have all been busy with sex. That does not mean that we *always* want to have sex that way. Some of us might be perfectly happy in the same room with twenty or thirty other people, all of them having sex, but sometimes we just want it to be private. Just the two of us, quiet, with no interruptions. It varies. It is 'of the moment'."

He thought about this and had to admit he could see the sense of it. "Yes, Golda, my dear Earth woman, you are correct. On my planet - hell, in my part of the galaxy! - It is considered rude to inquire about someone's sexual orientation. I see on this planet that you have many categories - the usual, and then the bizarre. On my planet, there is little that could be called bizarre. We have moved on

157

-we have grown up. Men and women, men and men, women and women... the list is almost endless. We have explored it all and absorbed it all. We take it on board on an individual case-by-case basis. There are factors, of course. We do not take at all kindly to anything that hurts others, which is why I mentioned my strength difference to you, back in the showers. I would not enjoy sex with you if there was the slightest possibility that I would damage you, even if you were one of those masochistic types who wanted to be hurt, or enjoyed pain. I might not just inflict pain on you; I really fear that I would kill you! That is unforgivable."

She smiled at him, this alien, this stranger in a strange land, and she had to admire him. He had travelled so very far to get to us, and she wondered how she could ever come up to his expectations, whatever they might be.

"Avvy, what did you just mean? You called me your 'dear Earth woman'. I like compliments as much as the next person, but what does that mean? Oh, yes, you and Ishmail had better have a talk after this. I read both of you like an open book. Your wording was slightly 'off' and I noticed it immediately." She stopped and directed her gaze to his eyes. "Tell me, Avraham, what does it mean in your language to call someone a 'good person'? I spotted Ishmail nearly having a fit when you said that."

He relaxed, back into the chair. Was there anything that this woman didn't see? He laughed. "Yes, Golda. My 'dear Earth woman'. Let me explain that one first. We have a large number of what you might call politically correct statements. These are the lines where you say something, but to those around you, those who know, you might have just said the opposite. They know the rules. Commander Ishmail and I have known each other for some years and, yes, we know the rules. I was deliberate in what I said, and he picked up on it straightaway, but he was too much the ambassador to let the secret out. He trusts me, and I trust him. He accepted what I said, and I was about to stand by."

She smiled at him, wanting him to say the right things. "And what is it that you did say, Avvy? Enlighten me."

He looked into her eyes, and she met his gaze. "Golda, I meant to say that I will see my days out with these people - these Israeli people. They are good, honest people who have no time for lies and deceit. I will see my days out with this small team of people - your flight crew, in particular and the rest of the Israeli's - as I trust them as much as I trust you, Commander Ishmail. Now, Commander Ishmail, deal with it!". He smiled at her, and even so, she was shocked, as the enormity of what he had just said came home to her - hard!

"Avvy - you said all of that? And you didn't think to tell us?"

"Oh, Golda. Did I need to? I needed my people to understand the situation - it could have gone wrong in so many ways. I needed them on my side, one hundred percent. Only then, when that was worked out, did I think I would ask you people whether you would agree. I had to get them first - only then could I even think of approaching you, and asking if you and I could work out some sort of agreement."

"Oh, Avvy, you didn't have to ask us - we knew - all of us knew from the moment that we were in the shower!" She was giving a huge smile, which Avvy liked to see.

"So we are good with this - all of this?"

"Avvy, we are *very* good with this - *all* of this!" She looked at him very warmly. "Avvy, check your instruments. Is it all OK with all of the readings?"

He did as he was asked and said, "They are all fine. The power is still building up, but it is looking perfect. Why?"

"Give Admiral Kaplinski an update. We need to see if we can sleep."

He did just that, finishing by saying that the sound system was going off, but they could be interrupted if it was necessary.

"Avvy, do you trust me? I trust you, but do you trust me?"

THE ADMIRAL

GETS CLOSE TO THE NATIVES

More than a little surprised by the question, he answered, "Yes, Golda, my Earth woman - I trust you. Why do you ask?"

"Good. Wait here." She left him and went to the gangway to the floor below. "Ben! Tsipi! Have you finished? Get yourselves up here, please."

There was a short interval, and then the two of them appeared, in their underwear, looking rather sleepy. "We were just settling down. What's up, Goldie?"

"I need you both to be my guardian angels." They both woke up. Did she just say...?

"Golda, we are yours. Say when. How is it going to happen?"

She looked at Avvy. "I don't know. I haven't asked him yet."

"What? Oh, come on..."

"Avvy, what you said, back in the showers, about being much stronger than us. Do you remember what we said in return? We said that we would look out for each other - we would watch over each other, even while one of us - me - was having sex with someone who is super-strong, and could crush me so very easily. Avvy, I want to try it. Now. These two will watch over us."

He was a little taken aback, but not too much. "Golda, my own dear Golda. I would be happy to try this now, with your dear friends - " he stopped. "With *our* dear friends looking out for your safety."

161

He stood up and, in the company of the other three, took all of his clothes off. He stood there, as he had before, in the shower at Dimona airport. Golda took a quick look, and fifteen seconds later, she was there, as naked as he was.

"Now, Avvy, do you remember what we discussed? If you were on top, then it might be risky, so I propose that you put your seat down and I lie on top of you. These others will hold your hands, if you let them, but I have to do most of the work."

"Agreed."

"Ready? Here we go, then," and she pushed him gently down onto his seat. It adjusted and became more like a bed. Golda was on top in seconds and did all of the work. Avraham struggled to stop himself from getting too carried away, but the hands-on approach from Ben and Tsipi kept him sane. He didn't kill her!

When it was finished, Golda had collapsed on top of Avvy, neither prepared to move, each enjoying the moment. It had been spectacular for the two of them, and the observers had a great time just seeing how an alien and an Earth woman could actually have sex!

It was Benyamin who said, "And now, I think that Tsipi and I will leave you, and we will go and have that sleep. You two enjoy each other's company a little while longer."

They left and went back to the bunk. They squeezed up and were both asleep in seconds.

Golda was exhausted. She fell asleep on top of Avvy even while he was busy looking at his displays, making sure that all of the balancing was correct and that the pods were performing correctly. While she slept, he brought up more displays, ones that told him what the total amount of junk and dust was, where it was positioned,

162

and where the strains on the pods were. It took half an hour, and he knew that he could now leave it to the computer system to do all of the work for the next few hours. Then he slept.

THE ADMIRAL

STARTS TO RELAX...AND THEN...

Avvy woke first and had to manoeuvre to get out from under Golda. She woke up with all the moving around.

"Hey, you. How are you? Did you sleep at all?"

"I did. I'm just checking how long we were out for, and then I need to check the current situation." He scanned through the displays and said, "We must have been tired. We were asleep for five hours! Wow, that's amazing."

"How are the pods doing?"

"I'm just checking them out... There looks to be no system failures, and I think that we need to see what the sky looks like."

At that moment, the other rejoined them. Ben and Tsipi were dressed, and Golda said, "Yes, we'd better put some clothes on, too. You don't know who might call!"

"Yes, OK, let's dress." That took a couple of minutes, and then Avvy said, "Time for us to take a look at the world. I have checked, and it is still there." With that, the sides of the pods' control cabin slipped away, revealing the open skies. They could immediately see that the screens were still going, pumping the dirt and dust high and away from Earth.

"That is an impressive sight, Avvy. I doubt I will ever see anything quite so stunning." Ben was in awe of what he was facing. Here he was, a human, in a space pod that belonged to an alien who had arrived in a disastrous way. Ben knew that the world was changing, even while they were talking, and seeing this most amazing sight.

"Avvy, we'd better check in. Ram and the Prime Minister will be going mad, wondering what we are up to."

"My thoughts exactly. Let's see how they have fared in the last few hours." He spoke into his radio, knowing that Ram would be there. "Ram, my Israeli friend. We are alive and well. How are things with you?"

They waited for a few seconds, and wondered what could have been the delay, and then Ram came on the radio. Greetings to all of you. In case you were wondering, even I have to attend to the call of Nature sometimes. Let me give you my updates, and then you can give me yours."

"We have had some attacks on both the land of Israel and attempts to intercept some of the pods. The South American pods are, as far as I can tell, all fine. Perhaps you need to verify that, Avvy? The troubles have been over here. We had some more arguing with Iran, and yes, Avvy, I had to instruct your scramjet pilots to do some hard work, but they were magnificent! They took down a flight of missiles and then destroyed the launching sites. I don't have any idea what Iranian casualty figures are, but at this stage of the game, I don't care. We have more important things to be talking about!"

"The Prime Minister has been dealing with a lot of interference from all over. There are countries that have suddenly decided that they must be kept up to date with these events - and yes, we have religious-based countries who say that we are going against God's will. We have kept all of the main countries that we got the U235 from in the loop, even though I think that they have done all that they can."

165

"What we have seen, over the last few hours, is that the screens that you are generating are definitely doing something, but owing to our lack of understanding of your technology level, we can't explain it to anyone - not very well, anyway - but it does look as though it is working! We have seen that the skies around the screens are still dark, but the 'push upwards' is definitely working - at least that's what we think. Can you confirm that, Avvy?"

He spoke while he was checking through the displays. "I can confirm that all of the pods are operational, and their fuel supplies are looking good. They should be working for a good few hours yet, even a day or more. This is why I started off really slow. It meant that they slowly built up the concentration. This is all to do with fuel economy. There is little difference in results between the South American pods and the ones over here - both theatres of operations are going well. I don't know that I can say anything about how long this will last, but it must be for at least another few hours - perhaps twelve or so. There has been a lot of work done, but there are still millions of tons of dirt and debris to get removed."

"Ram, can I speak to the Prime Minister?"

A familiar voice came back with "You already are, Admiral. I have been on a permanent loop since we first met. I just didn't want to get involved with everything, which is why I left a lot of the work to Ram and his very capable team. First of all, let me say 'thank you', Admiral, for your work. Whether we win or lose this fight, it will not be for want of trying. For that, I thank all of Israel, thank you. A lot of the rest of Humanity would also say 'thank you' if they could. It is the ones that want you dead that worry me."

Avvy looked at the others in the pod - *some people want me dead?* What is going on here? "Prime Minister - " and he was interrupted.

166

"Avvy, shall we dispense with the formalities? I am very tired and at this time I don't give a rat's arse who knows it." He sounded worn out, and Avvy guessed that he hadn't slept in days. "Avvy, we will definitely have to have a lot of drinks after this session, but by all that you call Holy, would you please call me Michael? If you are happy with Avvy, then I would like to keep calling you by that name. Agreed?"

Avvy smiled at the others in the pod. He knew these three so very well, and then there was Ram, and now Michael Rothburg, the Prime Minister of this tiny country. *How is it that from such a tiny country, you could get people of such massive character?*

He laughed, so that everyone could hear. "Michael, I would be honoured to call you by that name. Yes, you and I, and Ram, and Ben's whole flight crew are all invited to the party, wherever we manage to hold it - whether this works or not, we will have that damned party!"

"So, what next, Admiral? From what you have said, it is a waiting game. We cannot hurry this, but we need to be monitoring it very carefully, until it reaches a point that we can at least relax with."

"Michael, there is little that any of us can do, certainly for the next few hours. I would suggest that you have words with various other governments - start with those that are at least on your side - and tell them that there is much still to be done. The food and water chain damage is massive, and everywhere needs to be on full rations, starting now. Tomorrow is too late. This will require Martial Law all over the planet, and it means that you will have to come down very hard on people who ignore you. If you fail to establish the rules now, then it will make everyone else suffer over the next few months."

"I have been contacting most of the leaders in what we call the West. Most are horrified by the scale of the damage but agree that some action is better than none. We have talked briefly about the future, but you are right. We need to be putting things in place now; otherwise, we are wasting time. Ram, tell Avvy about the action in the Red Sea."

"Very well. You will need to verify everything, Avvy, but something, we don't know what, is moving through the Red Sea, up towards where you are. It doesn't show any signs of aggression, and I believe that it is connected, somehow, with all of the underwater activities that we have been seeing. Our latest information is that it is coming north and has just reached Mecca, or where Mecca was. It'll be with you in a few hours, given that it moves very slowly."

"Let me check out what I can find." He went back to the displays, pulling up new ones, closing down old ones. The others in the pod couldn't follow what he was doing, so they kept quiet and let him get on with it.

"Ram, I have a range of what you would call sensors. Various devices that allow me to see outside of the visible spectrum. I need to take this pod up and then move south over the Red Sea. I can't go too far, though, or I will start to interfere with the screens." He thought for a moment. "Actually, I'll take this pod up to get me a better view, but I will stay at this location. We have nothing else that we can do for some hours. Let us wait for this thing, whatever it is, to come to us."

Ram came back with "Whatever you say, Avvy. Keep a line open to us and report anything that you find."

"That I most certainly will do." He turned to the others in the pod. "OK, people. Time for another trip."

He made sure that he had kept the open view available, and the pod started to lift. The others were experienced fighter pilots, so there was no worry on their part about the new heights being shown. Within seconds, they were a few miles further up.

"Now, let's see what we can find out about whatever that thing in the Red Sea is. I could increase the power, and then this would register through the screens, but I don't want to go anywhere near them - not when we have done so much to get them operational. I will start looking just where the Red Sea meets the first part of the screen, and if I need to, then I will wait until the sensors start to register. This may take a while, so keep comfortable."

He watched the displays, every now and then playing with them. Nothing happened for at least half an hour, and then he, quietly, almost to himself, "Hello. What are you, then? Folks, I'll put this on the big screen." A new display showed up, taking over the whole of the centre of the pod. From their seats, they could all see it very clearly.

"Wow, I didn't even know that there was a big screen."

"There's a lot that you don't know about this pod. After this is over, we'll go through it, piece by piece. For now, let's just look at what it is in the Red Sea. My sensors are showing in your infrared, and from what I am looking at, this thing is very big and alive. It looks to have passed under the screens. The temperatures that are showing suggest that it is some form of live creature, although I've never seen one quite this big before."

"OK, so how fast is it, and how long before it gets here?"

"It's going really slow." It then dawned on him. "It's slow for a reason! It's huge, we know. If it went any faster, it would start creating tidal waves and swamp both sides of the Red Sea! This thing has to be intelligent!" He was now smiling, loving the idea of finding out what this thing was. "Oh, this is going to be exciting!"

"I don't know about you, but I have had a lot of excitement in my day so far. I can only take in so much," said Tsipi. "How long before it gets to this location?"

"At this speed, it'll be hours.

Silence from everyone, including those on the radio. Ben came in with "OK, there is much that we don't know about it, but what *do* we know about it?"

"It's big. I'd say perhaps three miles long, maybe more. It's quite wide, perhaps a few hundred feet, and from here I can't tell how high. It must weigh thousands of tons! How does something that size move...? It's alive, in some way, and I would suggest that it has an intelligence, although I don't know at what level. Is it controlling the other movements that have been reported around the world, the ones that have seen faults in the seabed being corrected?" He wondered. Could it be...? "Perhaps. If it is, then it is an intelligent life form that is miles ahead of us. Centuries or even millions of years ahead of us. People, we have to be very careful here. Michael, can we get anything from satellites, any visual details of the track of this thing? What's its history? Where did it come from?"

"We know that it started off the coast of Japan; we believe in what we know to be the deepest part of any ocean. Tracking back through the timeline shows that it first registered about two hours after you crashed into our world."

"Really? Nothing for centuries or more, and then this starts to show? That's weird. Anyone would think that when the ship crashed..." he trailed off, deep in thought, a lot of things flashing through his mind. *The ship is coming down, creating lots of chaos and noise. A great deal of noise, both above the water and below it. The shock waves must have gone everywhere, and it would have been incredibly...noisy!*

He fell back in his chair and laughed. His passengers wondered what had just happened, and Ram and Michael only heard the laughter.

The Prime Minister came on with "Avvy, I don't suppose that you'd like to share the joke?"

Still laughing, he managed to control himself and say, "It's not a joke. I think I have just realised something. It's rather strange, but try this on for size. If you are a child or an adult, and you are very sleepy, then sometimes you go to sleep and you can go into a deep, deep sleep. It means that you can sleep through noises around the house and noises from outside the house. You are gone, fast asleep to the world. Imagine then that you have been asleep for a few hours. You start to come out of your sleep pattern. Not immediately, but slowly. Imagine then that someone around the house slams a door or drops something that crashes onto a hard floor. There will be noise - a lot of noise! It will bring you out of your sleep abruptly. You will feel tired, perhaps upset, disorientated, that sort of thing."

They all looked at him, wondering where he was going with this, when he blurted out, "That's what we've done here! This thing, whatever it is, has been soundly asleep for years - probably thousands of years, and when my ship came in, it created a whole series of loud crashes. Not one but dozens of them, one after the other. There was no way that this thing could have slept through this! We woke it up. *I* woke it up!"

After a few moments of silence, where everyone tried to take in what had just been said, Ram came in with "That might work. In a weird way, just to add to an already weird day, that makes sense."

"What do we need to do about it, though? Have we annoyed this thing?"

"I don't know, Michael. I would say that initially it was probably peeved, but it has to be highly intelligent! It woke up, looked around, and saw the condition of the world. Somehow. I can't begin to guess how it works. It will be like me trying to explain the scramjet technology to a newborn baby, and expecting a serious answer! The connections may just not be there. Let's say that we did wake it up. It obviously is at home in the sea, so perhaps it was like that first cup of coffee in the morning, to set you up? After you have this coffee, then the world takes on a new look - you are calmer, better able to think straight, that sort of thing."

"It fits; sort of. It's no crazier than the rest of the day so far, so I'll buy it."

"But what do we do about it? *Can* we do anything about it, or *should* we do anything about it?"

"Like what? If this thing is anything close to what we think it is, then there is nothing that we can do, even if we thought that we needed to, which I don't, by the way. This thing, whatever it is, has shown every indication that it is working on our side, not against us. If it had wanted to, it could have taken us apart hours ago."

"Agreed, but what do we do now? Just wait for it to get to your position, and see what happens then?"

"Michael, I think that's exactly what we have to do. I can't do anything except monitor the pods. I can't move them, and I can't make them go any faster. I dare not change what is in place at the moment, so we have to wait for that to take its time. This strange shape will be some hours before it gets to us. We can monitor it along the way, but we have to just wait!"

"Avvy, I am not happy with it, but I agree. I don't like the not knowing. The idea that we have come so far, and we could be so close to getting the air problem sorted out, only to find that something else has happened to undermine all of that work."

"Michael, you will have to keep making the political noises that you have been making. Ram, keep me up to date with any information - especially if anyone launches again. I must monitor the pods for the next few hours, with as little disturbance as possible. All scramjet pilots, you are still with Admiral Kaplinski for the next few hours. I will tell you when I want to take back the control."

They all agreed with that statement. Avvy went back to looking at the pods, making sure that they were all functioning correctly. During the next few hours, Golda went below for a separate sleep, and then Tsipi came and joined her. Both were exhausted and slept soundly. Benyamin remained in the main chamber of the pod, aware that there was nothing he could do to help, but amazed that here he was, part of a small team trying everything to save what was left of humanity. *That's something to tell the kids, later on.*

Every hour or so, Avvy would supply an update. The fireworks outside were continuing, and Avvy was happy with the results that he could see. He contacted Ram, and between them, they discussed the Iranian airspace, the strange events in the Red Sea, and how the political world was facing up to a radically new future.

There were no further launches from anywhere, but the political aggression increased when previously strong countries faced up to the new reality - if they did not come on board with the changes, then they were to be left behind. All of the new world would be blocked to them. It was hard for the leaders in some countries to acknowledge this, but three-quarters of them eventually came around to the proposals.

Michael knew that the future battles would be with the remaining quarter of the countries, who seemed to be justifying what could only be called rampant ignorance. *Let's get through this set of battles first. I'll deal with this political nonsense tomorrow.*

After ten hours of further pod activity, Avvy voiced his thoughts to all. "I can still make no promises, because the job is not finished. It is, however, well advanced. All of the activity of the last few hours has gone amazingly well. I am delighted to report that a great amount of the pollution has been removed, pushed up into space. It may have played havoc with many of your communication satellites, but there is nothing I could have done about that. I did factor in the ISS, and I have moved around it when I had to. I am pleased to say that the ISS is fine, with no damage. Each country will have to assess its own satellites. There is a lot of clearing up still to do, and I think that these pods will run for another day, but I don't think I need to do anything else. It's all automatic. I would like your views, but I propose to run these pods for another few hours and then power them down. I have an idea that we need to see where the existing pollution has moved to - by then it will have travelled many miles, and covered large parts of the atmosphere. I think that I will have to trap each location and push the debris up out of the way on a one-to-one basis. This action here has been to remove eighty percent of the main problem, and right now, I think that it's worked!"

The other occupants started to clap, quietly at first, and then Ram and the others heard it and joined in. Avvy let them have their moment, and Golda came over and gave him a big kiss.

"Well done!"

"Thank you. Thank you all. It is quite true that I could not have done all of this without all of your help, so perhaps you should congratulate each other, too? Well done to each and every one of you."

"Avvy, you might want to check your sensors. That thing in the Red Sea is nearly in your position. Do you know what you plan to do?"

"Ram, I have no idea. Let me check." He went back to the displays, looking at various things that the others were still in the dark about. "Ram, I am going to drop our altitude. Can you keep everything focused on it for a while? I need to see how close I can get."

Although they didn't feel anything, they knew that the pod was dropping in its cruising altitude. It had to come down quite a way to allow a detailed, up-close inspection, if that was even possible.

After they had dropped for only a few moments, Ram came on the radio. "Avvy! Stop! Stop where you are. Something is happening. It's as if it's lifting out of the water. It's starting to show itself."

THE ADMIRAL

MEETS THE KRAKEN

He stopped the descent and told the others, "We are currently at a two-mile height. If Ram is correct, then this thing is coming out of the water. Let's have a look." The sides of the pod vanished, and they could see all around themselves.

Ben noticed it first and said, "Oh, wow. Will you look at that! It must be taking up a lot of space under the water."

"Ram, we see it. We are about two miles up, at a place called Ras Mohammed, which looks to be the most southerly part of this land. This thing, whatever it is, is huge. There is just a great mass of something coming out of the water. It looks to be just coming straight out." He looked again and wondered if it wasn't lifting up to meet them. It certainly looked like it. It just kept coming, out of the Red Sea, straight up, directly towards them. *This can't be a coincidence*, he thought! *This has to be deliberate!*

It continued to rise, achieving what looked to be an impossible height, and then stopped when it reached the same altitude that they were at.

Admiral Avraham looked at the shape that had appeared, his mouth open wide, his brain desperately trying to catch up with reality.

"Avvy, what the hell is that? It's huge!"

"I know. Let me think for a moment." He went silent, and he went over all of the stories that he had been told as a growing youngster. He looked at the shape, looming high and very wide across his view screens. "It has to be the Kraken! I believe that you have no end of those ancient stories, some tales from a time that is long gone. So do we. This creature that you have is exactly like our Kraken, a semi-mythological beast that we have put into our 'scary tales' category. It's the stuff to keep young children scared and afraid, not for adults. No one had actually seen one, at least not to talk of it." He looked at this absolutely magnificent beast, partly in disbelief and partly in full-blown admiration. *This is not a dream; this is real. Deal with it!*

He looked at his teammates. He regarded them very highly. It was hard to think of better people to face the future with. These were exceptional people. Would they rise to this level? He had no options – *I just have to see whether they will or not. We all sink or swim in the next few minutes.*

"People, there is much that is going on here that I do not understand. I will be honest with you - I have never experienced this before. What we are seeing here is not false. This is genuine. It is something from the darkest recesses of my past, and of yours, it seems. It is real. This is the stuff that dreams have been made of, and you and I here, today, must face it." He wondered what he could do next. How do you beat this kind of day - just how do you 'top' this?

"My friends, I must ask a great deal of you. You and I have just lived through what we thought may be our end. We all wondered if we would see into tomorrow. We have survived through that ordeal, and we are now into that 'tomorrow'. This -" and he looked at the still rising creature, wondering how the next few minutes may work out. "This is the final thing that we must face. This is the creature that makes or breaks a civilisation, or indeed a planet. What we have just lived through is nothing compared to what this creature has experienced, and I must ask that you trust me one more time – perhaps one last time. At this point, I do not know."

The others in the pod were stunned into silence. How do you answer that sort of thing? How does the 21st-century Man and woman desperately try to get their head around the idea that all that they had thought of their world, all that they had considered as a scientific marvel or breakthrough, all of that was as nothing, when compared to the might of this awesome beast that had appeared to the world? This was more than Godzilla, more than King Kong…this was more than anyone had ever considered, and it scared them. It terrified Benyamin, Golda, and Tsipi, and they all wondered if they were ever going to see their families again.

And then it hit home to one of them. Tsipi suddenly stopped in her doubts and her utter fear of what was in front of her. *Why fear this?* Why is it that you are in fear of something that could obviously have destroyed you and your people centuries ago, if not thousands of generations ago? Why are you so scared? And then she laughed. This caught all the others off guard. "How is it that you are laughing? We are facing death in the face, and you are laughing?"

"No, don't you see?" She looked at them and saw immediately that they did not see. "This is not death that we are facing, but *life*!" This was a higher form of creature that was as far removed from them as they were from the basic amoeba, and they didn't see it, but she did.

"Avvy, I think it is time that you and I trust each other for one more time. In the next few moments, you and I, and the whole world, will find out whether everything that we have survived through in the last few days was worth it." She rose from the seat and went to him. She offered her hand, and he, hesitating, took it.

"What are you talking about, Tsipi? There is so much that I –"

178

She stopped him and smiled at him. Softly, she whispered, "Avvy, this is it. This is where we make our mark on the world." She turned to the other two and said, "Disrobe. Completely, and join us at the pod's hatch." She turned back to Avvy. "Don't mess about – strip!"

He did so, in confusion, but not having a better option. All that he wondered about was 'where is this leading?'

Tsipi and Avvy walked towards the hatch and faced the open sky, with the utterly enormous beast taking up all of their view. "Avvy, how far can you extend the ramp? I would like it to go straight out for as far as it will go."

"It will go for perhaps a few yards. It is a mix of super-strong materials of all sorts. What do you plan? That we should walk – "

"Shhh." She turned affectionately to him. Softly, she said, "Yes, that is what we do. We walk out along the ramp. You and me, hand in hand." She called behind her. "Ben, are you ready? I need you to do the same behind us."

"We will be there, but I am not sure I have the faintest clue as to why!"

She wondered. Did she have the authority to do this? 'Oh, shut it!' screamed back at her! It's at that 'do it now or regret it forever after' moment.

She took his hand, this 'Avvy', this alien, this person who looked like a member of Humanity but clearly belonged to another branch of the evolutionary tree. They may resemble each other, physically, but the differences across the centuries of progress have shown themselves during the last few days. In the next few minutes, she could only hope that what they presented, as a common theme to Humanity, would allow this strange new creature to let them grow, to let them continue to expand their horizons, their science, their philosophy, their mathematics and their physics – their inquisitive nature, their search for those ever-elusive answers.

"Avvy, hold my hand, and just walk with me. If we are asked anything, and you think that you know how to answer, then do so. I may not know the language, not as well as you do. If you have any doubts, then say so at the time. This is not the time to hang back."

They both walked out along the ramp, the woman feeling reasonably sure of herself and the Man having major doubts about the sanity of this next operation. But he was a realist. Since he had no other viable option, he went with what was on offer.

The two of them walked slowly out across the ramp, and they heard that the other two had joined them. She forced the pace and continued to walk out, all of them wondering where this might lead. They all saw in front of them this strange creature, this wall that took out most of the sky. It was immense!

She got perhaps halfway across the ramp and then stopped. Avvy stopped too. There was a form of activity within the enormous creature. This is it. This is the moment. Sink or swim, do or die…

A thin strand came slowly out from the creature, headed in their direction. It was as if a tendril from the creature was being extended. It came out carefully, but very definitely in their direction.

"Stay your ground, people. Stay where you are."

The tendril was at the ramp and then stopped. They were not sure what was happening, but a shape at the end of it looked to be growing, a human shape. It only took a few moments, and then the shape detached from the tendril and stepped down to the ramp. It was female in form and looked directly at them. This strange new woman smiled. "Tsipi. Avvy. I am …" It looked to be hunting for a name, but gave up. "I do not have a name. I am here to talk to you. Will you listen?"

She couldn't believe what was going on, but managed to smile and say, "Yes, of course we will listen. We would love to listen. We understand that you have much to say, and we will listen to anything that you want to say to us."

"Good. Very good."

"Shall we return to the pod – our craft? It may be more comfortable there."

"An excellent idea. Do you think that I need clothes? I see that you don't bother, but do you think I need to?"

"With us, no, you do not need to worry about that. Perhaps when we meet other people, then we will all need to be wearing clothes, but that can wait."

They all walked back to the pod, Tsipi and Avvy hand in hand, the new stranger next to Tsipi. They got back into the pod, and the four of them wondered where the newcomer was going to sit.

"I must ask a few questions of you first. I need to have an understanding of this modern world, as I have not seen it for many years." She stood in the middle of the seats and looked around at each of them. "How do you people normally greet each other? Do you say 'Hello' or do you shake hands, or do you hold each other close, and have this 'sex' thing that I am finding out about?"

Golda coughed and said, "It can be any of them – it depends on the people, the location, the urgency – many different things. For instance, when we first met Avvy, we very quickly discovered that we all were very easy in each other's company, whether we had clothes on or not. He comes from another world, and one that is more socially advanced than we are here, on this planet." She smiled at that, and Avvy found himself smiling back, too.

"I understand that you have all been very busy. Well done, Avvy, on putting all of those pods of yours to work and preventing the air pollution from getting worse. I, in turn, have been very busy under the water, fixing a lot of broken areas of land that you call Tectonic Plates. Many were fractured, and I had to spend a lot of time and effort putting them carefully back together. So at the moment, I believe that we – you and I – have managed to put a halt to the damage. There is still much work to be done, in actually clearing up the mess that your ship caused, Avvy, but that can now happen in the future. Is now perhaps a good time for you to engage in that 'sex' that everyone has spoken about? I would like to see how this works, and perhaps join in myself, if you will permit?"

All four of them dropped their jaws at the same time. They looked at each other, then at the newcomer, and finally Golda said, "We are in the pod, at some high altitude. We have the place to ourselves, and it is unlikely that we will be disturbed for some time." She looked at each of them again. "Well, I'm game!"

Ben managed a "Me too!" and Tsipi said, "Oh, yes – Avvy?"

He was looking at the newcomer. "You do know that I am not like the others here?" She nodded. "I am many times stronger than any of them, and I don't know how my strength and yours will compare. My worry has always been that I may hurt someone when I engage in close sex. It is a physical act, and can be quite difficult to control, sometimes. I mention this so that you know I do not wish to harm anyone, including you."

She approached him and put her arms around his waist, pulling him close. "You worry, quite rightly, about these people, and how easily you could hurt them. Avvy, you do not have to worry about

182

me. You cannot harm me, even if you tried." She smiled and pulled him close against her. "I think you and I should make a start." With that, she gently pushed him down into his seat and promptly sat on his lap. "If you will excuse us, people, I think Avvy and I will be busy for a few minutes. Please feel free to amuse yourself." With that, she started.

The others looked on in surprise, for a moment, and then Tsipi said, "Ben, our turn. Golda, see what you can do to help them out." She manoeuvred Ben onto the seat and then sat astride him, and she started. Golda felt a little out of place, but then thought 'What the heck' and went over to Avvy. As much as she could, she joined in. When Avvy had finished, the stranger got off him and said to her, "Now it's your turn, Golda." This impossibly attractive non-human grabbed her, and they went down on the floor. By now, Ben and Tsipi had finished, and they sat up and watched as the two women started on each other. Avvy stayed seated and just watched. A random thought went through his mind *'What an impressive planet this is turning into!'*

When everyone had finished, they all stood up. What now? The stranger said, "People, in the past I have been known by many names, one of which is your Kraken." The others all looked at each other, and Avvy's eyes went wide. "As usual, the written accounts were all corrupted over time, which is how I ended up being painted as a 'bad guy'."

Avvy said, "On my planet, we have an equivalent of this 'Kraken'. It is a huge beast and has always been something to be feared. How is it that you are not what we expected? How is it that you are here and on my planet as well?"

"I am not the only Kraken. We are all one, even though we are physically spread across many galaxies. I do not actually know how

183

many of me there are. Do you know how many cells there are in your body?"

She continued, "My history is long – when I was last appearing, you had only started to write things down, and technology was at its earliest. You simply didn't have a clue what the world was, and no idea about the Universe at all! Please don't worry about it at the moment." She turned to Golda. "I see that you have taken to Avvy, and he has taken to you. That is as it should be. I should like to stay with you if I may, and see if I can assist with any further repairs. The damage has been stopped from continuing, but the damage is done. In a population of seven billion people, I believe that only four billion will survive, and the damage to the food chain is immense. If you are not very careful, then many more will die – perhaps half of the survivors – so there is a lot still to do."

"I believe for the moment that I need to set you people to be an even match. We already know that Avvy is old by your standards, but that is the normal way where he is from. I would ask you three to think on it – do not give me your answer now – but I am able to do many things. Things that you cannot even imagine." She turned to each of them in turn. "I like you, each of you. I like Avvy too." She smiled at him. "This world will need to have a clear set of organisers, people who can take charge, people who can get things done. We already know that Avvy has declared that the level of his technology is to be made available to you – and to Israel. Unfortunately, it seems that you have a lot of people in this world who, if they could, would steal this technology and use it for their own purposes. That must not be allowed to happen!"

Avvy voiced, "I agree, but how –" She held up her hand, stopping him.

"My offer to you is to make some of you people from this planet – and in particular you three here - more like Avvy. I can make you very strong, and very fast, and I can make you very old. My estimates would be that each of you – and Avvy – would easily make one thousand years."

They all caught their breath, including Avvy. One thousand? Most people didn't even make one hundred on planet Earth, even though on Homeworld, six or seven hundred was conventional.

Ben said, "One thousand years? You can do something that will see us perhaps get to one thousand years old? Why?"

"As I said, this world will need continuity for some time. It will need a steady hand, and I believe that you people are the steadiest that I could find." She smiled at them, and they were all wondering if this was real.

"Oh, wow," said Golda.

"But that raises a lot more questions, surely? What about our kids? We might start a family – do they inherit a really long life, or would they be normal? Do we stay as we are and watch the next few generations just grow old and die? That would hurt – a lot!"

"Ah, yes, this 'inheritance' thing. Your DNA is very complex, and I see that you are only just starting to understand it, and then perhaps bend it, and in places to mend it. What I am offering you is something that your science will most likely get to achieve sometime in the next few centuries anyway. You have achieved much, but you still have much more to achieve. Your children? Yes, they will inherit some of your DNA, but it will be under conventional science rules – some hit-and-miss logic where one of your children will live to be fifteen hundred years and another will barely make a century. That is the nature of this 'evolution' as you call it. I do not set the rules on this at all. While you think on this, I will say that while I am with you, the rest of me will slowly go back to our sleeping place."

Golda looked around at Ben and Tsipi. "I'm in!" she said, convinced that this was going to be an amazing life, one that she and her family could look forward to enjoying.

"Me too!" they both replied. How could they turn this down?

"Excellent," the stranger said. "Now, if I am to remain with you for a while, I suspect that I will need an Earth name. Does anyone have any suggestions?"

They all looked at each other, and Tsipi voiced, "Well, if you are a representation of the Kraken, then I might suggest something similar, perhaps 'Catherine'?"

She looked around, and they all agreed. "Good. Catherine. I like it! Well done, Tsipi. Now that we all know each other, I wonder if we shouldn't get back to base camp in Israel. Although I don't need much in the way of sleep, I think that you could all do with a good night's rest, but that will have to wait until you have updated the authorities there on what has happened. We should do that first, and then we can rest."

Ben said, "Agreed. I know that the Prime Minister will be chewing his arms off trying to figure out if we have won the war or not."

Avvy said, "Very well. We will put on some clothes and make our reports. Once that is done, I think we deserve a few days off – a holiday."

Ben could only suggest, "I don't know about a holiday. I intend to sleep for a week!"

Tsipi gave him a friendly shove. "I have no intention of letting you sleep for a week!" She turned to Catherine. "All that you were saying, about making us stronger and faster – when does that happen? Do we have to do anything?"

"Oh, it's done. You will now have to be very careful when you meet others on the planet, just as Avvy has been doing. He will tell you what to look for."

Benyamin summarised with "So, if I have this right, we are now super-strong, will live for a thousand years, and we are super-fast, too?"

Catherine could only answer "Correct."

Benyamin smiled. "Right now, this is what we do. We get back, make our reports, and then we need to vanish for a few days. We all have a lot to think about, and I don't think getting involved with the Israeli politics, the United Nations and any of the other countries and their internal squabbling will help. We have just saved this planet! We need a break! Avvy, set a course for home, please. Let's get this wrapped up."

"We are on our way. I don't propose to hurry, so we will be an hour or so."

Catherine said, "An hour? That's time enough to enjoy each other again. Ben, this time it's you and me. Golda, take Avvy and Tsipi, just enjoy the view!"

They all smiled. Life can be so very beautiful!

THE ADMIRAL

LOOKS FORWARD TO THE FUTURE OF HIS NEW HOME

They all went back to Dimona. They parked the main pod, and then Avvy parked all of the other spares. He gave one last look over the displays and was happy that they all looked to be operating correctly. They met up with Ram, and by now the Prime Minister had returned. Valerie and his Russian pilots were all there, and half of the scramjet fighters. Avvy explained that the others would be in the air, to the north and to the east of Israel.

"They are flying at an altitude of ten miles. There is little that Earth has that can touch them at such a height, but they can be down on your doorstep in seconds. Their instruments will see any further unauthorised launches."

Avvy and the people had all got dressed while they were returning, including Catherine, and he introduced her to everyone, and there were a lot of raised eyebrows when it was explained who she actually was.

It was the first time that the Homeworld crew had been seen by people from Earth. There was a lot of very careful congratulations from everyone, and it wasn't long before the party started and then, owing to everyone's exhaustion, ended. The local town had a few hotels, and Mossad had taken over everything that they could. The idea was to have the numbers split up across the different hotels for security reasons. Clearly, Mossad director Ram Kaplinski was still in charge, as he insisted that it was done that way.

188

"We will maintain high security over all of your pods, and I intend to maintain a very high security over each of you."

"Don't worry so much over the pods. They can look after themselves. Actually, I wonder if we shouldn't put them away somewhere, somewhere that nobody can get to them?"

"Well, that's an idea. They are very exposed at the moment. Where did you have in mind?"

"The bottom of the Galilee. I don't know how deep it is, but they will find their own balance. Any other ideas? I'll listen."

"No, I don't have any other ideas. Does this mean that you have to go to your pod?"

"Not at all. Hang on a minute." He went silent for a couple of minutes and then said, "OK, they will all launch and converge on the centre of the Galilee. They will quietly slip under the water and settle there for the night. I'll keep my pod in the hangar, though."

"Fair enough. Have you given any thought as to what we should do tomorrow?"

"Not yet, not really. I was going to get some rest after Golda and I had got to know each other a little better." He laughed, and so did she. By now, it was common knowledge that they both liked each other. "There is a lot to think on, but it will keep until we get a decent rest. The first that I've had in over a week!"

So it was that a few Israeli fighter pilots, an alien Admiral and just over six hundred of his people became the new face of the political world. Over the next few days, various scientists were contacted and brought over to stay in Tel Aviv. Many politicians were busy demanding that they be allowed in as well, but most were politely, at first, refused. The main countries that had helped Avvy

189

by supplying the U235 were all in daily contact with their counterparts in Israel, and a lot of close consultations were established.

The damage to the food chain was immense, and many thousands of square miles of crops and animals had simply vanished. The changes discussed by the different groups covered everything. Medicine was high on the list, as was food. There was no point in having alien medicine save millions of lives if they had all starved to death! Equally, there was no point in having millions of well-fed people who then succumbed to rampant diseases because of polluted water or food. The task was huge - immense! Most of the countries that had been hit were struggling, and Avvy had half of the pods re-designated as emergency relief transport. There was enough U235 still around for them to be active for a few weeks.

Israeli science was already a world leader. Over the next few days and weeks, it became *the* world leader. They were shown new formulas for disease control, new uses of existing materials that no one had yet identified. They had to wait all of three months before the first anti-cancer drugs were available and then tested on volunteers. Within another three months, they had their answer - it worked!

Other cures were identified - most blood-related illnesses were trapped and eliminated within the first year. A side effect of all of this curing of ailments was the prolonging of life expectancy. No one knew it at the time, but the introduction of the new medicines didn't just cure the ailment, but it also healed the rest of the body. Bone and tissue all mended, but it took three years before they realised what had happened, and when it was announced, then everyone wanted in on this. People have been living for over a century in their millions. They were content and working until they decided that they had had enough and wanted to retire, most of them well over eighty years of age.

The food chain was repaired, but it took time. The freshwater system had to be recreated in thousands of places, and millions more died while this was going on.

By the time that the new science had made itself effective, and people and the food chain had stabilised, around a total of three billion people had perished. Many had died on that first day, but then, owing to the rampant stupidity of some governments, more died. Some just refused outside help, claiming some insanity as 'It was God's will', or 'It goes against Nature'.

Some governments tried to be stupid and then found that they had a rebellion on their hands. It was usually short and very painful for all concerned. In many cases, Avvy had made direct approaches to various military leaders, bypassing the political element completely. This sometimes worked, as one soldier talking to another usually found a common ground, but not always.

It became necessary a few times for Avvy to take his pod fleet and invade. They tried every other option, and when it was doomed to fail then they had to create the change. The United Nations was dismantled, as it had proved its uselessness during all of this. A new 21st-century United Nations was established by the major helpers to Avvy, and it was based just outside of Jerusalem.

The Middle East went through a rough twenty-first-century education. Iran was effectively a wasteland, and the other countries around the area knew it. Iraq and Saudi Arabia were able to accept the situation, even though they did not move for change. Avvy and the others left them alone, and these countries were given no access to any of the new discoveries.

For those within Israel and its growing list of friends, things flourished. Within twenty years, the planet had technology and spacecraft that allowed it to explore the far reaches of the solar system. Within a century, the drive technology was developed that allowed them to go to the nearest stars and back within thirty years, but since three-quarters of the planet survived to be one hundred and fifty, this journey time was acceptable.

Man was truly setting itself up to be an explorer of the stars.

EPILOGUE

Yes, I plan for a second book. I think it deserves it. There are some interesting characters (well, I think so, and I wrote the book, so there!) and there are certainly no end of situations that they can find themselves in.

So, to Book Two!

Peter Keats

peterejkeats@gmail.com

www.ingramcontent.com/pod-product-compliance
Lightning Source LLC
Chambersburg PA
CBHW071604210326
41597CB00019B/3391